Jaguar XJ6

Series lll 1979-1986

Compiled by
R.M. Clarke

ISBN 1 85520 0864

Brooklands Books Ltd.
'Holmerise', Seven Hills Road,
Cobham, Surrey, England

Printed in Hong Kong

BROOKLANDS BOOKS

BROOKLANDS ROAD TEST SERIES
AC Ace & Aceca 1953-1983
Alfa Romeo Alfasud 1972-1984
Alfa Romeo Alfetta Coupes GT. GTV. GTV6 1974-1987
Alfa Romeo Giulia Berlinas 1962-1976
Alfa Romeo Giulia Coupes 1963-1976
Alfa Romeo Giulietta Gold Portfolio 1954-1965
Alfa Romeo Spider 1966-1990
Allard Gold Portfolio 1937-1958
Alvis Gold Portfolio 1919-1967
American Motors Muscle Cars 1966-1970
Armstrong Siddeley Gold Portfolio 1945-1960
Aston Martin Gold Portfolio 1972-1985
Austin Seven 1922-1982
Austin A30 & A35 1951-1962
Austin Healey 100 & 100/6 Gold Portfolio 1952-1959
Austin Healey 3000 Gold Portfolio 1959-1967
Austin Healey Sprite 1958-1971
Avanti 1962-1983
BMW Six Cylinder Coupes 1969-1975
BMW 1600 Col. 1 1966-1981
BMW 2002 1968-1976
Bristol Cars Gold Portfolio 1946-1985
Buick Automobiles 1947-1960
Buick Muscle Cars 1965-1970
Buick Riviera 1963-1978
Cadillac Automobiles 1949-1959
Cadillac Automobiles 1960-1969
Cadillac Eldorado 1967-1978
Chevrolet Camaro SS & Z28 1966-1973
Chevrolet Camaro & Z-28 1973-1981
High Performance Camaros 1982-1988
Camaro Muscle Cars 1966-1972
Chevrolet 1955-1957
Chevrolet Corvair 1959-1969
Chevrolet Impala & SS 1958-1971
Chevrolet Muscle Cars 1966-1971
Chevelle and SS 1964-1972
Chevy Blazer 1969-1981
Chevy EL Camino & SS 1959-1987
Chevy II Nova & SS 1962-1973
Chrysler 300 1955-1970
Citroen Traction Avant Gold Portfolio 1934-1957
Citroen DS & ID 1955-1975
Citroen SM 1970-1975
Citroen 2CV 1949-1988
Shelby Cobra Gold Portfolio 1962-1969
Cobras and Cobra Replicas Gold Portfolio 1962-1989
Cobras & Replicas 1962-1983
Chevrolet Corvette Gold Portfolio 1953 1962
Corvette Stingray Gold Portfolio 1963-1967
High Performance Corvettes 1983-1989
Daimler SP250 Sport & V-8250 Saloon Gold Portfolio 1959-1969
Datsun 240Z 1970-1973
Datsun 280Z & ZX 1975-1983
De Tomaso Collection No.1 1962-1981
Dodge Charger 1966-1974
Dodge Muscle Cars 1967-1970
Excalibur Collection No.1 1952-1981
Facel Vega 1954-1964
Ferrari Cars 1946-1956
Ferrari Dino 1965-1974
Ferrari Dino 308 1974-1979
Ferrari 308 & Mondial 1980-1984
Ferrari Collection No.1 1960-1970
Fiat-Bertone X1/9 1973-1988
Fiat Pininfarina 124 + 2000 Spider 1968-1985
Ford Automobiles 1949-1959
Ford Bronco 1966-1977
Ford Bronco 1978-1988
Ford Consul. Zephyr Zodiac MkI & II 1950-1962
Ford Cortina 1600E & GT 1967-1970
Ford Fairlane 1955-1970
Ford Falcon 1960-1970
Ford GT40 Gold Portfolio 1964-1987
Ford RS Escorts 1968-1980
Ford Zephyr Zodiac Executive MkIII & MkIV 1962-1971
High Performance Escorts Mk1 1968-1974
High Performance Escorts Mk II 1975-1980
High Performance Escorts 1980-1985
High Performance Escorts 1985-1990
High Performance Capris Gold Portfolio 1969-1987
High Performance Mustangs 1982-1988
Holden 1948-1962
Honda CRX 1983-1987
Hudson & Railton 1936-1940
Jaguar and SS Gold Portfolio 1931-1951
Jaguar XK120 XK140 XK150 Gold Portfolio 1948-1960
Jaguar MkVII VIII IX X 420 Gold Portfolio 1950-1970
Jaguar Cars 1961-1964
Jaguar Mk2 1959-1969
Jaguar E-Type Gold Portfolio 1961-1971
Jaguar E-Type 1966-1971
Jaguar E-Type V-12 1971-1975
Jaguar XJ12 XJ5.3 V12 Glold Portfolio 1972-1990
Jaguar XJ6 Series II 1973-1979
Jaguar XJ6 Series III 1979-1986
Jaguar XJS Gold Portfolio 1975-1988
Jeep CJ5 & CJ6 1960-1976
Jeep CJ5 & CJ7 1976-1986
Jensen Cars 1946-1967
Jensen Cars 1967-1979
Jensen Interceptor Gold Portfolio 1966-1986
Jensen Healey 1972-1976
Lamborghini Cars 1964-1970
Lamborghini Cars 1970-1975
Lamborghini Countach Col No.1 1971-1982
Lamborghini Countach & Urraco 1974-1980
Lamborghini Countach & Jalpa 1980-1985
Lancia Stratos 1972-1985
Land Rover 1948-1973 - A Collection
Land Rover Series I 1948-1958
Land Rover Series II & IIa 1958-1971
Land Rover Series III 1971-1985
Land Rover 90 & 110 1983-1989
Lincoln Gold Portfolio 1949-1960
Lincoln Continental 1961-1969
Lotus and Caterham Seven Gold Portfolio 1957-1989
Lotus Cortina Gold Portfolio 1963-1970
Lotus Elan Gold Portfolio 1962-1974
Lotus Elan Collection No.2 1963-1972
Lotus Elite 1957-1964
Lotus Elite & Eclat 1974-1982
Lotus Turbo Esprit 1980-1986
Lotus Europa Collection No.1 1966-1974
Marcos Cars 1960-1988
Maserati 1965-1970
Maserati 1970-1975
Mazda RX-7 Collection No.1 1978-1981
Mercedes 190 & 300SL 1954-1963
Mercedes 230/250/280SL 1963-1971
Mercedes Benz SLs & SLCs Gold Portfolio 1971-1989
Mercedes Bens Cars 1949-1954
Mercedes Bens Cars 1954-1957
Mercedes Bens Cars 1957-1961
Mercedes Bens Competion Cars 1950-1957
Mercury Muscle Cars 1966-1971
Metropolitan 1954-1962
MG TC 1945-1949
MG TD 1949-1953
MG TF 1953-1955
MG Cars 1959-1962
MGA & Twin Cam Gold Portfolio 1955-1962
MGA Roadsters 1955-1962
MGA Collection No.1 1955-1982
MGB MGC & V8 Gold Portfolio 1962-1980
MGB Roadsters 1962-1980
MGB GT 1965-1980
MG Midget 1961-1980
Mini Cooper Gold Portfolio 1961-1971
Mini Moke 1964-1989
Mini Muscle Cars 1961-1979
Mopar Muscle Cars 1964-1967
Mopar Muscle Cars 1968-1971
Morgan Three-Wheeler Gold Portfolio 1910-1952
Morgan Cars 1960-1970
Morgan Cars Gold Portfolio 1968-1989
Morris Minor Collection No.1
Mustang Muscle Cars 1967-1971
Oldsmobile Automobiles 1955-1963
Old's Cutlass & 4-4-2 1964-1972
Oldsmobile Muscle Cars 1964-1971
Oldsmobile Toronado 1966-1978
Opel GT 1968-1973
Packard Gold Portfolio 1946-1958
Pantera Gold Portfolio 1970-1989
Plymouth Barracuda 1964-1974
Plymouth Muscle Cars 1966-1971
Pontiac Tempest & GTO 1961-1965
Pontiac GTO 1964-1970
Pontiac Firebird 1967-1973
Pontiac Firebird and Trans-Am 1973-1981
High Performance Firebirds 1982-1988
Pontiac Fiero 1984-1988
Pontiac Muscle Cars 1966-1972
Porsche 356 1952-1965
Porsche Cars in the 60's
Porsche Cars 1960-1964
Porsche Cars 1964-1968
Porsche Cars 1968-1972
Porsche Cars 1972-1975
Porsche Turbo Collection No.1 1975-1980
Porsche 911 1965-1969
Porsche 911 1970-1972
Porsche 911 1973-1977
Porsche 911 Carrera 1973-1977
Porsche 911 Turbo 1975-1984
Porsche 911 SC 1978-1983
Porsche 914 Gold Portfolio 1969-1976
Porsche 914 Collection No.1 1969-1983
Porsche 924 Gold Portfolio 1975-1988
Porsche 928 1977-1989
Porsche 944 1981-1985
Range Rover Gold Portfolio 1970-1988
Reliant Scimitar 1964-1986
Riley 11/2 & 21/2 Litre Gold Portfolio 1945-1955
Rolls Royce Silver Cloud 1955-1965
Rolls Royce Silver Shadow 1965-1981
Rover P4 1949-1959
Rover P4 1955-1964
Rover 3 & 3.5 Litre Gold Portfolio 1958-1973
Rover 2000 + 2200 1963-1977
Rover 3500 1968-1977
Rover 3500 & Vitesse 1976-1986
Saab Sonett Collection No.1 1966-1974
Saab Turbo 1976-1983
Shelby Mustang Muscle Cars 1965-1970
Stubebaker Gold Portfolio 1947-1966
Stubebaker Hawks & Larks 1955-1963
Sunbeam Tiger & Alpine Gold Portfolio 1959-1967
Thunderbird 1955-1957
Thunderbird 1958-1963
Thunderbird 1964-1976
Toyota Land Cruiser 1956-1984
Toyota MR2 1984-1988
Triumph 2000. 2.5. 2500 1963-1977
Triumph GT6 1966-1974
Triumph Spitfire 1962-1980
Triumph Spitfire Col No.1 1962-1982
Triumph Stag 1970-1980
Triumph Stag Collection No.1 1970-1984
Triumph TR2 & TR3 1952-60
Triumph TR4-TR5-TR250 1961-1968
Triumph TR6 1969-1976
Triumph TR6 Collection No.1 1969-1983
Triumph TR7 & TR8 1975-1982
Triumph Herald 1959-1971
Triumph Vitesse 1962-1971
TVR Gold Portfolio 1959-1990
Volkswagen Cars 1936-1956
VW Beetle Collection No.1 1970-1982
VW Golf GTi 1976-1986
VW Karmann Ghia 1955-1982
VW Kubelwagen 1940-1975
VW Scirocco 1974-1981
VW Bus. Camper. Van 1954-1967
VW Bus. Camper. Van 1968-1979
VW Bus. Camper. Van 1979-1989
Volvo 120 1956-1970
Volvo 1800 1960-1973

BROOKLANDS ROAD & TRACK SERIES
Road & Track on Alfa Romeo 1949-1963
Road & Track on Alfa Romeo 1964-1970
Road & Track on Alfa Romeo 1971-1976
Road & Track on Alfa Romeo 1977-1989
Road & Track on Aston Martin 1962-1990
Road & Track on Auburn Cord and Duesenburg 1952-1984
Road & Track on Audi & Auto Union 1952-1980
Road & Track on Audi 1980-1986
Road & Track on Austin Healey 1953-1970
Road & Track on BMW Cars 1966-1974
Road & Track on BMW Cars 1975-1978
Road & Track on BMW Cars 1979-1983
Road & Track on Cobra, Shelby & GT40 1962-1983
Road & Track on Corvette 1953-1967
Road & Track on Corvette 1968-1982
Road & Track on Corvette 1982-1986
Road & Track on Datsun Z 1970-1983
Road & Track on Ferrari 1950-1968
Road & Track on Ferrari 1968-1974
Road & Track on Ferrari 1975-1981
Road & Track on Ferrari 1981-1984
Road & Track on Fiat Sports Cars 1968-1987
Road & Track on Jaguar 1950-1960
Road & Track on Jaguar 1961-1968
Road & Track on Jaguar 1968-1974
Road & Track on Jaguar 1974-1982
Road & Track on Jaguar 1983-1989
Road & Track on Lamborghini 1964-1985
Road & Track on Lotus 1972-1981
Road & Track on Maserati 1952-1974
Road & Track on Maserati 1975-1983
Road & Track on Mazda RX7 1978-1986
Road & Track on Mercedes 1952-1962
Road & Track on Mercedes 1963-1970
Road & Track on Mercedes 1971-1979
Road & Track on Mercedes 1980-1987
Road & Track on MG Sports Cars 1949-1961
Road & Track on MG Sprots Cars 1962-1980
Road & Track on Mustang 1964-1977
Road & Track on Nissan 300-ZX & Turbo 1984-1989
Road & Track on Peugeot 1955-1986
Road & Track on Pontiac 1960-1983
Road & Track on Porsche 1961-1967
Road & Track on Porsche 1968-1971
Road & Track on Porsche 1972-1975
Road & Track on Porsche 1975-1978
Road & Track on Porsche 1979-1982
Road & Track on Porsche 1982-1985
Road & Track on Porsche 1985-1988
Road & Track on Rolls Royce & B'ley 1950-1965
Road & Track on Rolls Royce & B'ley 1966-1984
Road & Track on Saab 1955-1985
Road & Track on Toyota Sports & GT Cars 1966-1984
Road & Track on Triumph Sports Cars 1953-1967
Road & Track on Triumph Sports Cars 1967-1974
Road & Track on Triumph Sports Cars 1974-1982
Road & Track on Volkswagen 1951-1968
Road & Track on Volkswagen 1968-1978
Road & Track on Volkswagen 1978-1985
Road & Track on Volvo 1957-1974
Road & Track on Volvo 1975-1985
Road & Track - Henry Manney at Large and Abroad

BROOKLANDS CAR AND DRIVER SERIES
Car and Driver on BMW 1955-1977
Car and Driver on BMW 1977-1985
Car and Driver on Cobra, Shelby & Ford GT 40 1963-1984
Car and Driver on Corvette 1956-1967
Car and Driver on Corvette 1968-1977
Car and Driver on Corvette 1978-1982
Car and Driver on Corvette 1983-1988
Car and Driver on Datsun Z 1600 & 2000 1966-1984
Car and Driver on Ferrari 1955-1962
Car and Driver on Ferrari 1963-1975
Car and Driver on Ferrari 1976-1983
Car and Driver on Mopar 1956-1967
Car and Driver on Mopar 1968-1975
Car and Driver on Mustang 1964-1972
Car and Driver on Pontiac 1961-1975
Car and Driver on Porsche 1955-1962
Car and Driver on Porsche 1963-1970
Car and Driver on Porsche 1970-1976
Car and Driver on Porsche 1977-1981
Car and Driver on Porsche 1982-1986
Car and Driver on Saab 1956-1985
Car and Driver on Volvo 1955-1986

BROOKLANDS PRACTICAL CLASSICS SERIES
PC on Austin A40 Restoration
PC on Land Rover Restoration
PC on Metalworking in Restoration
PC on Midget/Sprite Restoration
PC on Mini Cooper Restoration
PC on MGB Restoration
PC on Morris Minor Restoration
PC on Sunbeam Rapier Restoration
PC on Triumph Herald/Vitesse
PC on Triumph Spitfire Restoration
PC on VW Beetle Restoration
PC on 1930s Car Restoration

BROOKLANDS MOTOR & THOROUGHBRED & CLASSIC CAR SERIES
Motor & T & CC on Ferrari 1966-1976
Motor & T & CC on Ferrari 1976-1984
Motor & T & CC on Lotus 1979-1983

BROOKLANDS MILITARY VEHICLES SERIES
Allied Mil. Vehicles No.1 1942-1945
Allied Mil. Vehicles No.2 1941-1946
Dodge Mil. Vehicles Col. 1 1940-1945
Military Jeeps 1941-1945
Off Road Jeeps 1944-1971
Hail to the Jeep
US Military Vehicles 1941-1945
US Army Military Vehicles WW2-TM9-2800

BROOKLANDS HOT ROD RESTORATION SERIES
Auto Restoration Tips & Techniques
Basic Bodywork Tips & Techniques
Basic Painting Tips & Techniques
Camaro Restoration Tips & Techniques
Custom Painting Tips & Techniques
Engine Swapping Tips & Techniques
How to Build a Street Rod
Mustang Restoration Tips & Techniques
Performance Tuning - Chevrolets of the '60s
Performance Tuning - Ford of the '60s
Performance Tuning - Mopars of the '60s
Performance Tuning - Pontiacs of the '60s

BROOKLANDS BOOKS

CONTENTS

ACKNOWLEDGEMENTS

This is the latest of Brooklands Books' many volumes devoted to Jaguars and their Daimler-badged equivalents, and in it we cover the six-cylinder XJ models built between 1979 and the original XJ6's demise in 1986. Enthusiasts of the 12-cylinder cars may wish to know that we have recently covered these in one of our Gold Portfolio books, entitled Jaguar XJ12, XJ5.3, V12, 1972-1990.

Brooklands Books are an archive service for today's owners of the interesting cars of the past, and their aim is to make available again road tests and other technical pieces which were written about the cars when they were new. For their very existence, they depend upon the generosity and understanding of those who hold the copyright to the articles they reproduce, and our thanks on this occasion go to the managements of Autocar, Autosport, Car and Driver, Car – South Africa, Fast Lane, Modern Motor, Motor, Motor Manual, Motor Sport, Performance Car, Road & Track, Road Test and Wheels. We are grateful also to motoring author James Taylor, who has written the short introduction which follows.

R.M. Clarke

The third incarnation of Jaguar's XJ range appeared in March 1979, rather more than ten years after the launch of the original cars in September 1968. Three engines were available, as had been the case with the now-superseded Series 2 cars, and these are 3.4-litre and 4.2-litre versions of the six-cylinder XK engine, plus the 5.3-litre V12. Once again, Daimler-badged versions were also available with all three engines.

Many enthusiasts regretted the passing of the "pure" XJ6 shape which graced the Series 1 and Series 2 cars, but it was undoubtedly true that the design needed a facelift after such a long time in production. It was also true that the result of the extensive changes made under the skin of the Series 3 was a quieter, faster, less thirsty and better-equipped XJ range.

The Series 3 XJs were born at a low point in Jaguar's fortunes, when poor quality control – which affected all the British Leyland companies at the time – had seriously damaged sales in the important US market. But a year later, John Egan was appointed to the Chairmanship of the company, and he tackled the problems so effectively that Jaguar had regained most of its old respect and recaptured most of its lost sales within a couple of years.

There were range realignments in 1982 and 1983, when the Daimler name was dropped altogether in Europe and was then made more exclusive in other markets, but it was by now no secret that a new XJ6 was on the way. Coded XJ40 inside the Jaguar company, it eventually emerged in 1986 and replaced the six-cylinder XJs, although the V12 cars remained in production. It was perhaps the best single tribute to the excellence of the superseded XJ6 that the new models were so much like them.

The XJ Jaguars inspired enthusiasm from the moment they were launched, and they will continue to do so for many years yet. This book is a very welcome addition to the bookshelf of anyone who believes, as I do, that the XJ Jaguars were some of the greatest luxury cars of the modern era.

James Taylor

At last the new Jaguars are here. A 10 car range of much-improved models that should fend off the continental opposition well into the '80s — the . . .

Facelifted felines

Above: at a glance, the Series III Jaguars look much as before, but a lengthier look reveals less cluttered styling and a crisper roof line

Left: the beautiful six-cylinder engine now has Lucas-Bosch electronic fuel injection which gives it more power and torque yet consumes less fuel

IT IS difficult to believe that the XJ6 and the XJ12 were introduced over 10 years ago, for these elegant cars do not appear dated today. Towards the end of 1974, however, Jaguar decided that the car could be improved by increasing the glass area, and by raising the roof line about an inch to provide both more head room for the rear seat passengers and enable a sliding roof to be fitted without encrouching on the head room of the front seat passengers. It was also felt that the appearance would be improved by increasing the rake of the relatively upright windscreen pillars. The problem was carrying out these changes above the waste line without compromising the very successful styling of the car by producing a top half that was no longer compatible with the lower half.

It is the profile of the Series III Jaguars that has changed the most. Increasing the rake of the windscreen pillars by three inches and thereby eliminating the front quarter lights together with the sharper roof line and the deeper side windows has given the car an altogether leaner look when side on. The increased glass area at the side has been obtained by raising the roof line slightly so that the rear seat passengers now have more head room, and by narrowing the roof and thereby increasing the inwards "tumble home" or curvature of the side windows from waistline to roof. The curvature of the backlight on the other hand has been flattened slightly without any ill effect on the size of the rear parcel shelf.

Mechanically, though, B-L's face-lifted range of prestige cars has changed little. There are no major changes at all to the mechanical specifications of the XJ12 Jaguars and Daimlers or the XJ6 3.4, but both the 4.2 litre XJ6 and the six cylinder Daimler Sovereign of the same capacity are fitted with a Lucas-Bosch electronic fuel injection systems thereby coming into line with XJ6 4.2s exported to the United States since May of last year. This and other modifications to the 4.2 litre engine have increased its maximum power output from 170 bhp at 4,500 rpm to 200 bhp at 5,000 rpm while maximum torque of 236 lb.ft is now developed at only 2,750 rpm instead of 3,000 rpm, thereby making the car more flexible to drive. Improved fuel economy is also claimed for the fuel injected engines.

The Lucas-Bosch fuel injection is similar to the system that has been fitted to the XJ12 since 1975, so there is plenty of in-service experience with it on Jaguars. However, the sensor in the air intake manifold that feeds information to the electronic control unit now measures the air flow in the manifold and not the air pressure as on the V12, as this has been found to be much more critical than the precise timing of its injection into the

Above: new bumpers, a more sloping A pillar and thinner C pillar give the Series 111 car a subtley altered appearance.
Below: Daimlers retain their ornate grille

cylinder, which now takes place in two squirts instead of a single squirt. Fuel pressure has also been increased from 28 psi on the V12s to 36 psi, and bimetal strip instead of a wax capsule thermostat now cuts out the special cold start injector when the engine attains its running temperature.

The five-speed manual gearbox announced at the 1978 Motor Show and based on the Rover box is now available for all the six cylinder Series 111 Jaguars and Daimlers with the exception of the Daimler Vanden Plas 4.2 for which the automatic transmission is standard. As it is on all the V12 Jaguars and Daimlers. The five speed box weighs 30lb less than the four speed plus overdrive transmission previously fitted, and although the final drive ratio is now 3.31:1, mph per 1000 rpm in fifth gear remains 27.5 mph, owing to the different ratios of the new box, as shown in the accompanying table.

Another modification attracting attention is the new massive black wrap-round bumper design arrangement at front and rear — black mouldings with bright cappings. Although the new bumpers are now very similar to those fitted to cars bound for the United States, they do not have the 5 mph impact beams demanded by US Federal regulations. The front bumpers contain the indicator lights and the rear bumpers the fog guard lights — except on US specification cars. The sidelights have been removed from their former location beneath the outer

SERIES III 4.2 TRANSMISSION TABLE

	Series III 5-speed	Series II 4-speed	Series III 3-speed automatic
1st	3.321	3.238	2.400
2nd	2.087	1.905	1.460
3rd	1.396	1.389	1.000
4th	1.000	1.000	
5th	0.833	0.779 (o/d)	
Final drive	3.31	3.54	3.31
Mph/1000 rpm	27.5	27.5	22.9

headlamps, where they would now be obscured by the new bumper and are incorporated in the headlamps. The radiator grille of the Jaguar models has abandoned the horizontal theme introduced for the Series II range in 1973 and has reverted to vertical bars and a central rib reminiscent of the style of the Series I Jaguars. The Jaguar's head radiator badge is gold on bronze for the XJ12 5.3 and gold on black for the 4.2 and 3.4 XJ6 models, but the 5.3 litre cars no longer sport a V12 motif beneath the Jaguar badge. The laminated windscreen is now fitted by thermal adhesion so that it in effect forms part of the body structure, thereby increasing its torsional rigidity as well as reducing the chance of stray water leaks. The rear window is fitted by the same method. Tinted glass is standard on all models except the 3.4 where it is an optional extra. No changes have been made to the radiator grille of the Daimler models.

The only other way of telling a 12 from a six at the front is by the leaping Jaguar motif mounted low down just behind the front wheels, for it is in gold for the XJ12 and silver for the XJ6s. Much more eye catching from a side view are the new wheel trims in black and stainless steel which follow the very sensible modern fashion of leaving the wheel nuts exposed instead of concealed behind a hub cap which is often difficult to remove should a wheel change become necessary.

The door handles are now flush fitting and have lift and pull operation instead of the former push buttons. The rear has been tidied up somewhat, the new rear lamp cluster now incorporates the reversing light and gives a larger area of illumination. The number plate housing is now flatter and wider, extending the width of the number plate, and has a boot lift latch similar to the one fitted to the XJ-S. And when overtaken by a Jaguar, the script on the right of the number plate will tell you whether it was an XJ6 or XJ12.

The finish of the Series III cars should be better than that of any previous Jaguars, for the bodies are now being painted in a brand new £15 million plant located at Castle Bromwich where the bodies are put together. Moreover, this new plant applies four coats of thermoplastic acrylic paint which is noted for retaining its gloss over a long life.

Many new refinements have been added to the Series III cars either as standard equipment or as optional extras. The Associated Engineering Econocruise cruise control, so useful for maintaining a set speed for mile after mile on motorways without the driver's throttle foot going to sleep, is now available as an optional extra on the automatic transmission 5.3 and 4.2 models. Other new optional equipment includes a steel sun roof operated by a

Left: restyled seats feature adjustable lumbar support, deeper backrests and repositioned headrestraints

Above: Daimler interior still has traditional appeal but now comes with remote-control door mirrors, leather covered, restyled steering wheel and interior light "delay"

Right: Jaguar dash unfortunately is little changed and the instruments are still marred by reflections

double cable from an electric motor in the boot and controlled by a console switch, wash-wipe for the outer headlamps, fitted as standard on the Daimler Vanden Plas, as are also twin electric remote control door mirrors which are optional extras on all the other Series III cars.

Standard equipment now includes intermittent as well as flick operation for the windscreen wipers, giving a six second delay between each sweep of the blades which now give a bigger final sweep when switched off in order to park automatically out of sight: There is also a new three-key security system, with an ignition key, a master key and a service key that locks and unlocks only the doors and the petrol filler cap and so can be left with a garage without giving access to the boot or the glove box.

Quartz halogen headlights replace the previously fitted tungsten sealed beam units on all save the 3.4, mono radio and stereo cassette players are fitted to all models except the Daimler Vanden Plas 4.2 and V12 which has Stereo VHF radio reception.

Modifications to the interior include two additional warning lights in the instrument panel centre strips and indicate low coolant and fog guard lamps switched on, while a new bulb failure warning light in the speedometer face lights up should a side light or stop light bulb fail.

Perhaps the most appreciated interior change will be the new front seats with seat backs one and a half inches higher for extra support and with a new lumbar feature that enables the driver or front seat passenger to vary the amount of lumbar support bulge in the squab by turning a control knob until the curvature of the squab fits into their back. Electrically controlled front seat height adjustment is standard on both Daimler Vanden Plas front seats and is an optional extra for the driver's seat only on all other Series 111 models.

Another nice touch is the inclusion of a brief case shaped moulded plastic toolkit complete with handle and clipped to the boot side wall.

Driving impressions

It can happen that even relatively minor modifications can spoil what was previously a very good car, but I was relieved to find when I drove the new Series III cars in the West Country recently that they have definitely been improved. Relieved, because without being so dogmatic as to state that the XJ12 is the finest saloon in the world, the Series II was definitely the finest I have driven. Until the Series III which thanks to its increased glass area, more comfortable seats, new deep pile carpets and other refinements is certainly a pleasanter car to enter. On the move, the differences between a Series II and a Series III V12 seemed minimal. Slightly more feel to the power steering, perhaps, which seemed to hiss more than previously, a very quiet engine, a very good automatic transmission with swift-acting kickdown and that glorious feeling of almost limitless power under one's right foot. But a very docile big cat which would purr softly through a village then soar up to its high cruising speed with one long sustained push from the seat squab.

Curiously, the extraction of more power from the 4.2 seems to have widened the gap between it and the V12 rather than lessened it, for the six now seems more sporting in character. The engine had more of a throaty roar when accelerating hard than I remembered, and with its five speed manual transmission one was inclined to drive it in more sporting fashion, rather than just rely on the sheer power of the V12. The gearbox was somewhat stiff, but then the mileometer recorded only 1,800 miles and I know from my Rover that these gearboxes do not loosen up until around 7,000 miles have been covered. A rough check of the car's instruments showed it to be cruising at 2,500 rpm/70 mph in fifth and at 2,950 at the same speed in fourth. On both XJ12 and XJ6, alas, reflections from the instruments remain a problem.

Although more sporting in feel, the 4.2 has gained in flexibility from the switch to fuel injection, to such an extent that a five speed manual transmission seemed rather superfluous.

Philip Turner

Perfecting the near-perfect?....

. . . well, Browns Lane try to: Jaguar's new Series III XJ range described and driven by Michael Scarlett

Photographs by Peter Cramer

THEY AREN'T perfect. There are design and manufacturing weaknesses. They have problems and failures, as does every other manufacturer. But compared with their international rivals, Jaguar's XJ saloons are unrivalled. They are nothing to be complacent about — but all the same, as we said to Bob Knight, chief architect of the original chassis, and now Jaguar's managing director, is it not quite a responsibility to change the XJ greatly?

''It scares me to death'', said Mr Knight. What needs changing? Remembering *Autotest* quibbles, we would straight away list our dislike of the crude-feeling gear selector on the automatic and absurdly old-fashioned windscreen wiper movements; other weaknesses include no seat height adjustment and an incomplete central locking system (which doesn't lock the boot). Somebody has criticised lack of rear headroom apparently — and there are people who have dismissed the styling as old-fashioned and therefore automatically poor. That last point of view is not *Autocar*'s, the XJ shape is not one that conforms to 1979 fashions, but, pleasing as some of today's styling trends for once are, that doesn't self-evidently mean that yesterday's style is bad, especially when, as in the Jaguar's case, it is (in our opinion) such a near-perfectly proportioned example. Besides good, solid if boring mass production reasons like keeping tooling costs down, Jaguar have been sensible not to fall for the transatlantic panic to make major facelifts every four years in a foolish Gadarene rush to pay tribute to the false god fashion. The XJ body is nearing 11 years old — and is still, to our eyes, superbly handsome. What have they dared do to it?

At first sight, you may say ''Not much'', adding perhaps ''thank goodness.'' Then something slightly odd will strike your eye, as it might on the approach of a familiar face underneath a slightly too short, angular haircut. There are in fact a lot of changes, mostly details — including our short list. But you were right about the haircut.

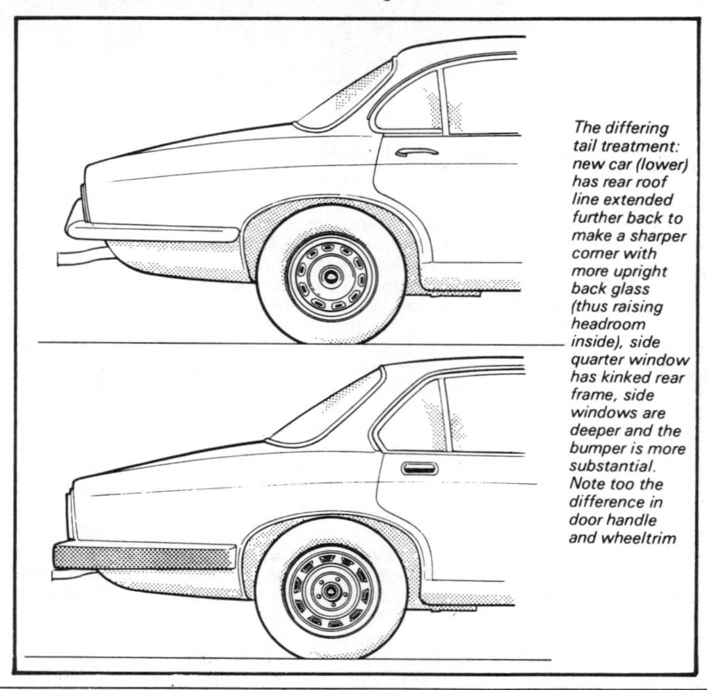

The differing tail treatment: new car (lower) has rear roof line extended further back to make a sharper corner with more upright back glass (thus raising headroom inside), side quarter window has kinked rear frame, side windows are deeper and the bumper is more substantial. Note too the difference in door handle and wheeltrim

Body

The height of the body is the same, but in order to give more headroom at the back, the roof line has been raised slightly as it nears the rear, reducing the inclination of the back window and giving that slight sharpening you half-noticed on that first look. In front, on the other hand, the windscreen pillars are more heavily raked (three inches more, measured at the roof end). The roof width has been reduced slightly, giving more tumblehome of the sides (from roof to waist). Windows are deeper and the front quarter light has been deleted to give the car something nearer the enviable cleanness of looks of the lamented two-door coupé. All models have laminated screens, and these are now thermal-glued in place, giving a claimed improvement in bodyshell stiffness.

An electric sunshine roof, a modified Inalfa one, is now optional. Its operating rack runs down the inside of the rear quarter pillars and across the front side of the squab panel; the drive motor is on the other side of the panel to cut down noise intrusion.

Below the waist, door handles are now flush fitting and there are new bumpers which make a tidier job for Jaguar of producing cars to suit Europe and America. European cars have a black injection moulded centre to give a little protection from tactile parkers, but for America this section is replaced by a more projecting one. US 5 mph protection

Main picture: Three views of the new Jaguar/Daimler Series III range. The white car is a Daimler Sovereign 4.2, the yellow one a Jaguar XJ6 4.2 and the red one a Jaguar XJ12 5.3. Distinguishing features are the restyled front and rear ends, the deeper windscreen and rear window and the raised roofline at the rear. Visible on the XJ 4.2 is the electrically-operated sunshine roof which is offered as an optional extra across the Series III range. New front and rear bumpers are covered by black injection mouldings

Bottom left: New quartz halogen headlamps (extra on the XJ6 3.4) now incorporate the sidelamps while indicators are recessed in the bumper. Wash/wipe for the headlamps is an optional extra on all models except the Daimler Vanden Plas, on which it is standard

Bottom right: Revised rear clusters now incorporate reversing lamps. Fog rearguard lamps are built into the bumpers. There is new boot badging too, with the XJ6 nomenclature making a comeback

Above: Restyled XJ6 3.4. Note the deeper side windows and the recessed door handles. The boot release is now positioned adjacent to the rear number plate lamps. Above right distinctive wheeltrims in black and stainless steel are a Series III feature. XJ6 3.4 models have an all-black hub and a black centre around the wheel nuts, 4.2 models have a stainless hub and black centre, while 12-cylinder models have all-stainless trims. This is a Jaguar XJ6 4.2

Right: Under the bonnet of the XJ6 4.2 carburettors have been replaced by electronic fuel injection, a system standard on the V12 models for some time

Below left: The driving compartment of the Series III Jaguar has many changes. Most noticeable are the steering wheel, new cloth trim (standard on the 3.4 pictured here), standard fitment of a Philips radio and stereo cassette player, and symbols instead of words on switches

Below right: Height of the front seat back has been raised by 1½ in. and the head restraints are now positioned nearer to the head. Carpets are more luxurious and there is a radio speaker in each door

Perfecting the near-perfect? . . . *continued*

is as before, by Menasco telescoping struts, but the reinforcing beam backing the bumper for the US is now an aluminium alloy extrusion, which saves a total of 35lb front and rear. Front bumpers contain the indicators and the rear ones carry the fog guard lamps (not on US cars) inside, instead of bracketed below.

In front the Daimler grille is unchanged, but the Jaguar one is slightly different, with vertical bars and a centre rib instead of the previous grid. Sidelamps, separate on the Series II, are combined with the outer headlamp now. There are bigger area back lamps, and number plate lamp and boot handle are altered. Twin door mirrors are fitted, manual for Britain and electric (extra here) for Europe. Cars fitted with the standard wheels have different wheeltrims.

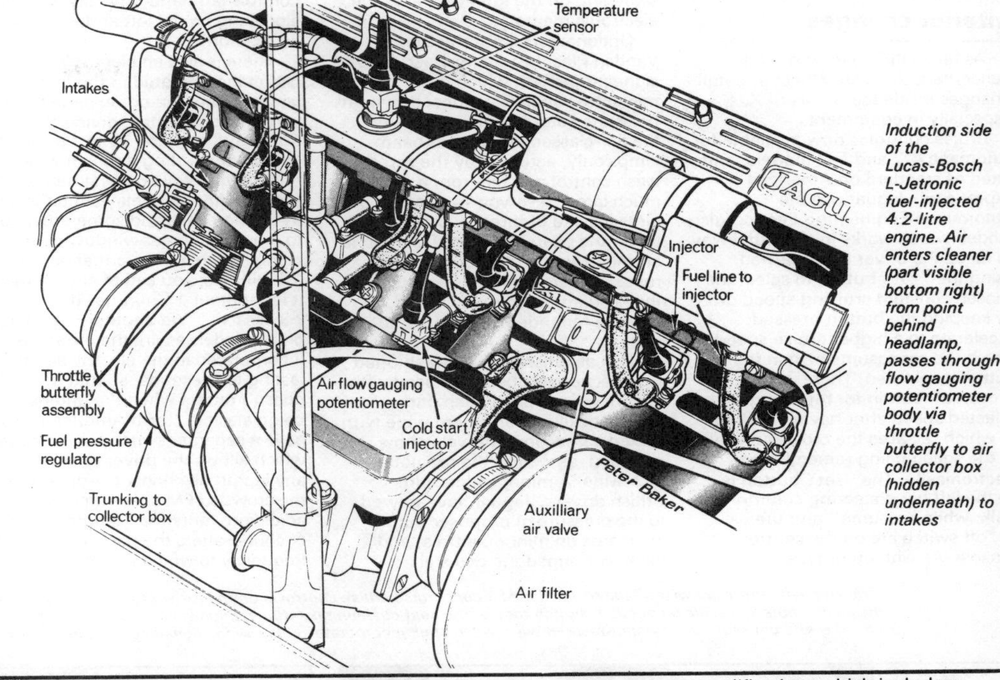

Induction side of the Lucas-Bosch L-Jetronic fuel-injected 4.2-litre engine. Air enters cleaner (part visible bottom right) from point behind headlamp, passes through flow gauging potentiometer body via throttle butterfly to air collector box (hidden underneath) to intakes

Fuel injection for the 4.2

This represents the most exciting improvement for the majority of XJ customers, who account for around two thirds of Jaguar's market. The adoption of Lucas-Bosch L-Jetronic electronic fuel injection is combined with a considerably higher compression ratio — now 8.7 instead of the previously modest 7.8-to-1 — slightly larger inlet valves — 1⅞in. instead of 1¾in. — and earlier inlet opening (22deg before top dead centre instead of 17deg). The higher compression happens because of the use of the old E-type "9-to-1" ratio pistons and the extra space taken by the bigger valves and in spite of a new, thicker gasket. The result is a very worthwhile increase in maximum power, which goes up by 19 per cent from 172 bhp (DIN) at 4,700 rpm to 205 at 5,000. Maximum torque goes up from 222 to a considerable 236 lb ft but occurs at a higher speed — 3,750 instead of 3,000 rpm. Jaguar say that overall efficiency has improved noticeably, so that economy is appreciably better. This is not simply an improvement in day-to-day overall consumption due to better cold start economy obtained automatically thanks to the stricter control exercised by a good injection system, but applies apparently throughout the regime. On over-run, the fuel supply to the injectors is cut off completely as a further economy and of course in the interests of better emissions; it cuts in again at 1,200 tpm. Other gains include less maintenance (Jaguar-Daimler franchise holders will be equipped to deal with the system even though it is different from the 12-cylinder car's Lucas-Bosch D-Jetronic set-up) and less engine noise thanks to the elimination of the hot air flap for the air cleaner of carburettor cars, which produced some intake roar.

A fuel injection 4.2 XJ has in fact existed for the past year; the car began production in this form for North America, meeting Federal emission standards with the addition of Bosch Lambda feedback control catalyst but without using an air pump or exhaust gas recirculation. In this shape it has an 8.1 compression ratio for running on 91 octane low-lead fuel and develops 177 bhp at 4,750 rpm.

The 3.4 continues for Europe only with a pair of SUs, and the V12 is also unchanged.

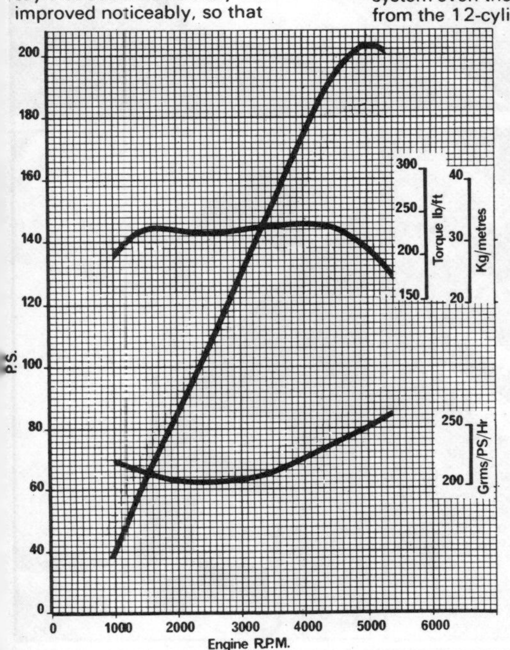

Power (Din — in metric bhp; 1 PS=0.986318 hp), torque and specific fuel consumption curves for the 4.2 fuel-injection engine — extra performance and efficiency gained with better mixture control at all times, a higher compression, larger inlet valves and earlier inlet opening

That 77mm box

Jaguar said that they were going to offer the Rover SD1 77mm overdrive five-speed gearbox in place of the previous Jaguar four-speed and separate electric Laycock de Normanville overdrive last year, but in fact only now is it to be made available as the manual option on 4.2 and 3.4 models. The delay seems to have been due to noise problems, acceptable in the quiet Rover, but not in the extraordinarily quiet Jaguar. As fitted, the 77mm box for the XJ has modifications which include a stronger layshaft, bigger bearings, and a needle roller bearing instead of a plain one for reverse — these modifications incidentally applying to Rovers as well. Actually the needle roller reverse had been adopted for police Rovers already, according to Jaguar's engineering director of power units and transmissions, Harry Munday, "for reversing 40 miles up M1."

That elusive five-speed manual box — a superb gearchange

Most regrettably, it is held that too few manual gearbox 12-cylinder customers exist for there to be any possibility of putting Harry's 84mm (the dimension of the shaft centres) overdrive five-speed into production. Very similar to the 77 mm — which it antedates — it has the right ratios and a superb gearchange and, of course, more than enough strength for the 5.3. Failure to produce it means the end of any manual box XJ-S after the last of the four-speed gearbox stock runs out in the very near future — and of Jaguar's right to claim to make the world's only 150 mph high production Grand Touring car. A great pity — claiming the world's only high-production 142 mph GT may be just as practical, if more challengeable, but it isn't the same.

The automatic boxes remain the same — Borg Warner's Model 65 for the six-cylinder cars and GM's "Turbo-Hydramatic" for the twelves — although the Model 65 will in due course give way to the Model 66 which we are assured will bring with it that long wanted improvement in selector change quality. We hope that it will also include a stop to prevent inadvertent selection of neutral from drive and the deletion of the present unnecessary stop between drive and 2.

continued

Interior changes

As far as the driver is directly concerned, there are a host of detail changes inside the Series III XJs, especially in equipment.

Cruise control is now available for automatic 4.2 and 5.3 models, and fitted as standard on American export cars. Jaguar use the Automotive Engineering Econocruise model, which works in the usual way as far as the driver is concerned; using the "set" button to select the chosen constant cruising speed or, by keeping the button pressed, accelerate to a higher cruise speed which the car assumes when the button is released. There is a "resume" button for returning to the selected speed after having braked — which releases the cruise control — the speed being remembered electronically. The "set" button is on the left hand steering column stalk, while "resume" and the on/off switch are on the centre console. As with other types, switching off the ignition clears the electronic memory.

Optional on all but the Daimler Vanden Plas, headlamp wash/wipe is the proper sort which uses miniature wipers as opposed to high pressure water jets. They work on the flat-glassed outer main beam lamps only, actuated by the screen wash control and fed from a very much larger reservoir — nearly 12½ pints instead of three — placed under the nearside wing. The feed is arranged so that the screen wash runs out first, leaving 1½ pints for the headlamps; this way round, the driver is reminded of the need to refill.

That astonishingly old-fashioned wiper action has been replaced with a more normal one which parks automatically; intermittent wipe with a fixed six-second interval is now provided, as well as the previous flick-wipe. A minor sadness for British drivers, Jaguar have bowed to the pressure of proposed future standards on minor controls, and have rearranged the two stalk controls left-hand-drive style, with signalling on the left and wipe/wash on the right.

There is also an electric delay circuit incorporated in the interior lamp circuit — in our experience a very welcome detail, giving the occupants an extra 10 to 15 sec of light after shutting doors in which to find ignition key and seat belt sockets. Another delay circuit is used as on Mercedes and some other cars to prevent the rear window heater being left on for more than around 15 minutes; you can of course re-set it for another 15 minutes if necessary. Yet another, standard on all models as are the others, is incorporated in the electric aerial, lasting ten seconds, so that it doesn't retract when the engine is being started. North American cars have a separate switch for the aerial, which allows the driver to set it to any length he chooses. Apparently in the crowded FM spectrum there, adjusting "antenna" length (which of course alters the particular frequency to which it is most receptive) is a favourite gambit of the American driver eager to improve selectivity — in other words, to make sure he is tuned into one particular station rather than two.

All Series III XJs for Britain will have radio/stereo cassette players as standard, using the Philips AC 460 AM mono radio combination on all but Daimler Vanden Plas cars which have the AC 860 FM stereo radio one. Four speakers are provided, one in each door.

The other considerable electrical change is the adoption of powered cushion tilt adjustment as an option for the driver's seat only on all but VDP cars which have it on both front seats. It is not proper height control — the arrangement moves the back of the cushion through an arc of nearly two inches, pivoting about the front; it will be interesting to see how useful this arrangement is. Control is via a rocker switch on one corner of the seat base. Lumbar support in front can now be varied by up to 1½in., by turning a knob at the inside of each seat.

From top left, anti-clockwise: Bellows for the AE Econocruise cruise control is on the exhaust side of the engine; speedometer markings have changed — note kph scale on outside; switch for electric seat cushion rake (fire extinguisher is not standard); lumbar support adjuster knob is on inside of each front seat; new symbol-marked wipe-wash stalk incorporates delay wipe; signalling and dip stalk is now left-hand-drive mounted, on left

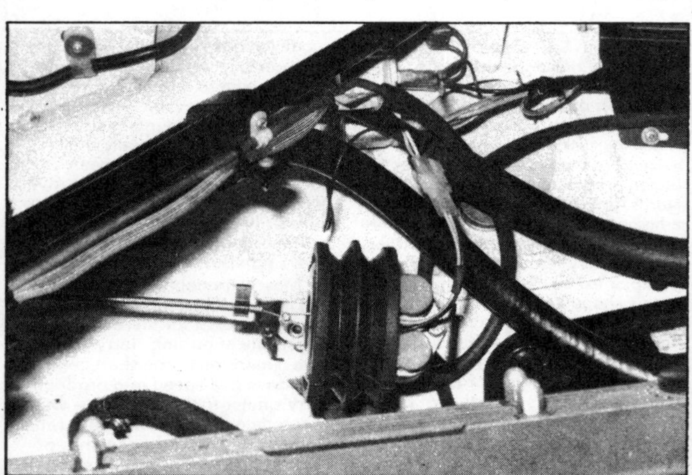

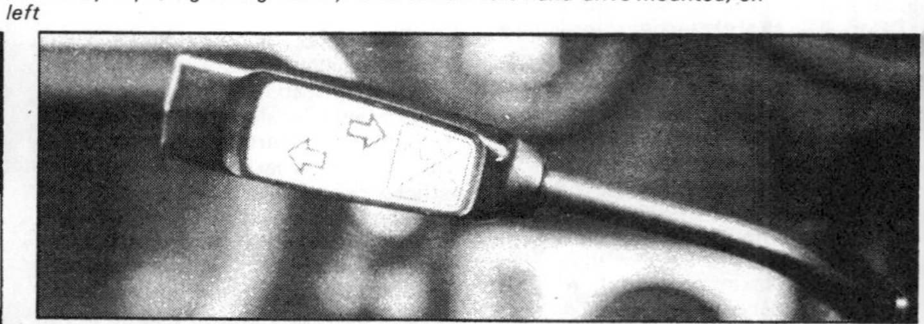

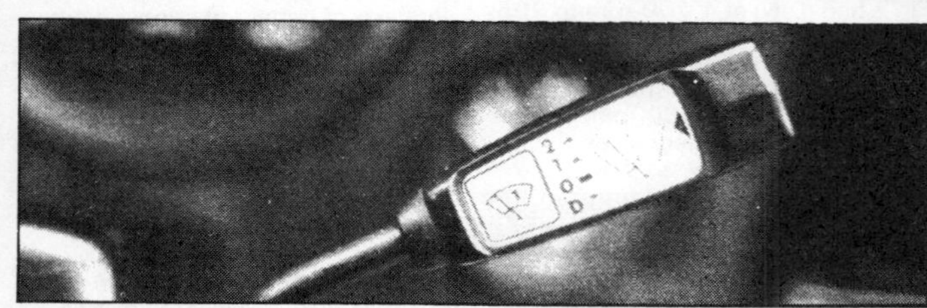

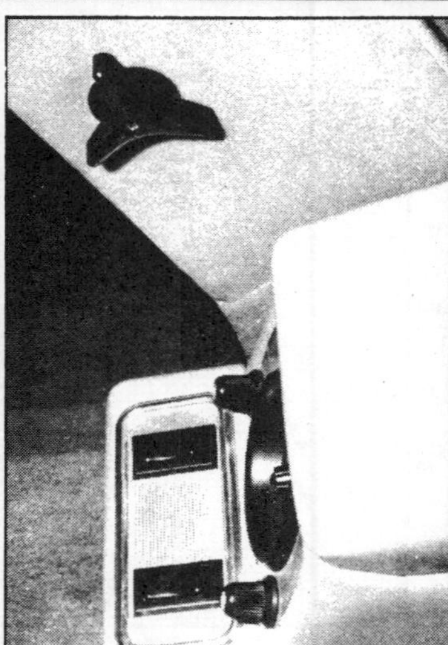

Perfecting the near-perfect? . . .

Series III

Door mirrors with manual interior control are standard on both sides, except on VDP and US specification cars which offer electric ones, optional on the rest. The optional electric sunshine roof mentioned earlier has an interesting detail found necessary in development. When open, in spite of a wind deflector, it suffered from the pneumatic flutter effect encountered at speed on many cars so fitte . Someone eventually tried holding a cigarette packet as an extra deflector at the centre of the roof opening, and this completely eliminated the effect. That is the explanation for the small cigarette-packet-shaped rectangular addition to the deflector.

There are detail changes for the electro-magnetic central locking system, which now locks the boot — but doesn't unlock it (Jaguar believe that owners who get back into the car would prefer this, so that no one can get at the boot whilst you are stationary in, say, a traffic jam) — and which now is worked from either the driver lock itself or the interior locking lever (instead of using the rocker switch on the centre console. It still doesn't control the fuel filler as the Mercedes one does. There are three keys — a master one which deals with everything except the ignition for which there is a special key, and a ''service'' key, which leaves the boot and glove compartment locked if you so wish.

There are now warning lamps to cope with low coolant — a good thing since we know of one case at least of an owner who ruined an engine unnecessarily through failing to realise that the fan belt had broken — rear fog guard and bulb failure covering side lamps and brake lamps (the latter whether a bulb or the handbrake or stop lamp warning switches fail). All except XJ 3.4 models now have halogen headlamps instead of the more expensive-to-replace tungsten sealed filament type.

By using vacuum-formed rubber and foam insulation for door panels, bulkhead and propellor shaft tunnel, Jaguar claim to have improved still further the already exemplary interior quietness of the XJ. The minor annoyance of power steering hisses still present on the Series III XJs we tried should be absent on production cars thanks to the introduction of an extra sound shield between the interior and the rack.

continued

From below left, anti-clockwise: That improved headroom and the bigger side windows are obvious here; back of each front seat has map pocket; boot dimensions are unchanged but still generous; sun visors are recessed into headlining; sunshine roof with power operation is available

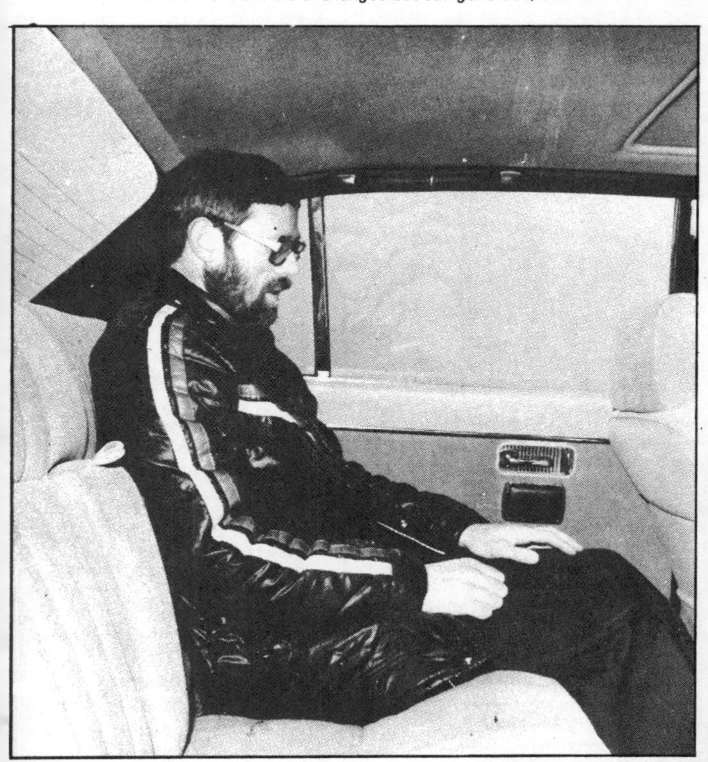

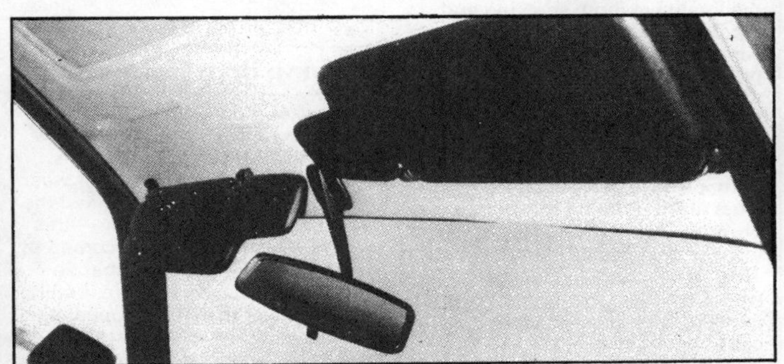

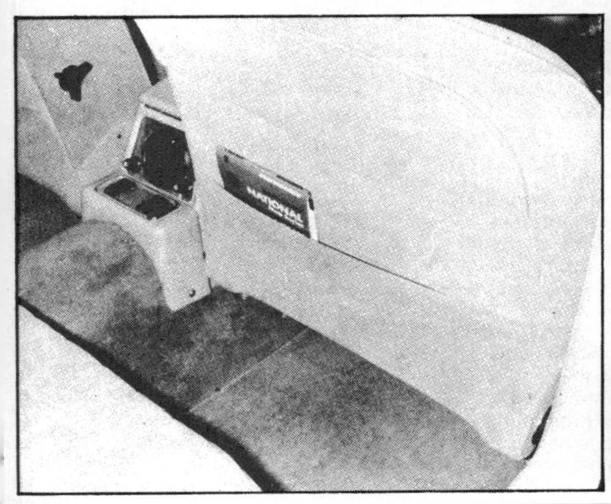

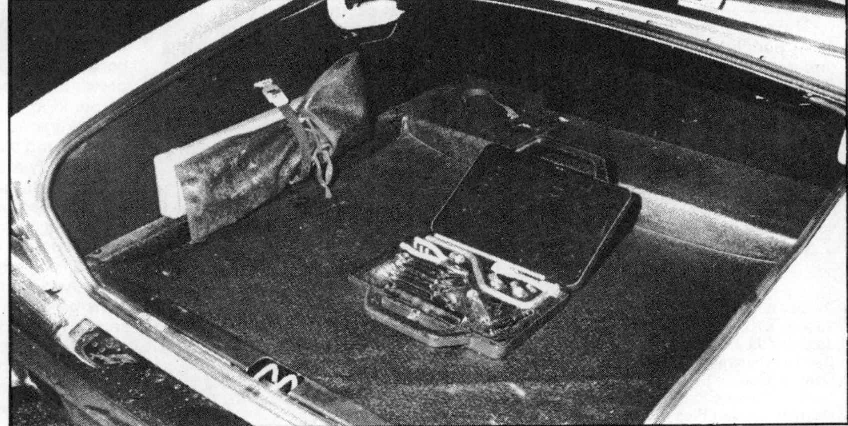

Perfecting the near-perfect? . . .

Jaguar toolkits used to be one of the minor delights of the car, encased neatly in a box that fitted very tidily inside the spare wheel. For some time this high standard was dropped, sadly, but it seems that things may be looking up with the introduction of a moulded plastic toolcase containing wrench, plug and other spanners, pliers, screwdriver and spare bulbs and fuses which is clipped to one side of the boot.

A welcome return to former standards is the neatly cased and comprehensive toolkit, complete with spare bulbs and a tyre pressure gauge

New paint plant

As well as a new range of colours, Jaguar will have the benefit of a new £15.5 million Pressed Steel Fisher paint plant just commissioned at Castle Bromwich, claimed to be the latest and most technically advanced in Britain. Built in the first place to paint the Series III cars, it is sited alongside the body build area itself. Processes include phosphate pre-treatment (both spraying and dipping), electro-priming, primer surfacing (two coats) an ''adhesion promoter'', four coats of thermoplastic acrylic colour aided by hand application to critical areas, oil sanding and ''re-flowing'' where the body is passed through an oven for 20 minutes to give a durable high gloss finish. Added protection is provided with undersealing plus wax injection into enclosed box sections. There is a random inspection whereby certain bodies are taken aside for a ten-day salt-spray corrosion resistance test.

Prices

The new range costs an average of 11 per cent more, with the XJ 4.2 understandably the most additionally costly, at 12.1 per cent more, thanks to the fact that it is the most extensively revised in specification. Prices range as follows; notice one unpleasant fact — for the first time there is a Jaguar model — strictly a Daimler — which costs more than £20,000.

	New price	Old price	Per cent increase
Jaguar XJ6 3.4	£11,188.71	£10,338	8.2
Jaguar XJ6 4.2	£12,325.95	£10,994	12.1
Jaguar XJ12 5.3	£15,014.61	£13,430	11.8
Daimler Sovereign 4.2	£12,983.49	£11,646	11.5
Daimler Double Six	£15,688.53	£14,096	11.3
Daimler Vanden Plas 4.2	£17,208.36	£15,516	10.9
Daimler Vanden Plas 5.3	£20,277.27	£18,219	11.3

EQUIPMENT TABLE

	Jaguar			Daimler		Daimler Vanden Plas	
	XJ6 3.4	XJ6 4.2	XJ12 5.5	Sovereign 4.2	Double Six 5.3	4.2	Double Six
Automatic transmission	O	O	S	O	S	S	S
Manual 5-speed ttransmission	O	O	—	O	—	—	—
Air condcitioning	E	E	E	E	E	S	S
Tinted glass	E	S	S	S	S	S	S
Sunroof	E	E	E	E	E	E	E
Philips mono radio / stereo cassette	S	S	S	S	S	—	—
Philips stereo radio / cassette	E	E	E	E	E	S	S
Electric radio aerial	S	S	S	S	S	S	S
Four radio speakers	S	S	S	S	S	S	S
Cloth trim	S	O	O	O	O	—	—
Leather trim	E	O	O	O	O	S	S
Head restraints	S	S	S	S	S	S	S
Inertia reel rear seat belts	E	E	E	E	E	S	S
Alloy wheels	E	E	E	E	E	—	—
Twin fog lamps	E	E	E	E	E	S	S
Quartz hologen headlamps	E	S	S	S	S	S	S
Electric door mirrors (2)	E	E	E	E	E	S	S
Adjustable reading lamps	—	—	—	—	—	S	S
Headlamp wash / wipe	E	E	E	E	E	S	S
Front seat electric height adjustment	E	E	E	E	E	S	S
Cruise control	—	E	E	E	E	E	E

E= *Extra-cost option,* O= *Option at no extra cost,* S= *Standard equipment. A dash indicates that an item* **is** *not available on that model*

Driving impressions

It takes more than the generous morning we spent driving the new range to evaluate improvements properly when the original car was so generally excellent. We liked the lumbar support adjustment — this works well and will be welcomed by every driver — it is a pity that on such a prestigious car there is still nothing finer than the stepped rake adjustment for the seat back. Other minor cribs concern the very slight gear whine we encountered on the 4.2 manual car in second gear, plus, on the 5.3 a small but perceptible final drive note. A tiny point which is amusing rather than annoying is a surprising consequence of the extra tumblehome of the body sides — if you drive along a roadside which is relatively dark (like one edged with a high hedge), there is a curious reflection which suggests to the corner of the eye that there is a car running alongside you. In fact it is the less steeply leaning side glasses reflecting the bright chromium-plated ashtrays in the centre console.

The manual car was of course the most interesting. It seems very promising, with the usual Rover superb gearchange and wonderfully long legs at 27.55 mph per 1,000 rpm (near enough the same as the old overdrive car's 27.5). The biggest difference is surely that extra top end power and the even more extraordinary flexibility, which almost approaches that of the legendary manual XJ-S — you can accelerate from well below 500 rpm in top without protest from that wonderful six. An interesting consequence of the added top end power is perhaps less fortunate — you tend to stray more often into the upper regions of the engine's regime, and although there is appreciably more power there, the old slight roughness of that engine has not been banished, so that things become a little stressful. All the same, one is reminded that the original conception of the XJ6 was of a sports four-seater — and this transformation certainly restores a sporting tinge to the car's still magnificently untroubled demeanour. The great thing is that they don't seem to have spoiled the car — it *is* certainly even better than before. □

SPECIFICATION

ENGINE	Front, rear drive
Head / block	Al Alloy / cast iron
Cylinders	6, in line: dry liners
Main bearings	7
Cooling	Water
Fan	Viscous
Bore, mm (in)	92.07 (3.62)
Stroke, mm (in.)	106.00 (4.17)
Capacity, cc (in^3)	4,235 (258.4)
Valve gear	Dohc
Camshaft drive	Chain
Compression ratio	8.7-to-1
Ignition	Electronic
Fuel injection	Lucas-Bosch L-Jetronic Electronic
Max power	205 bhp (DIN) at 5,000 rpm
Max torque	236 lb ft at 3,750 rpm
TRANSMISSION	
Type	5-speed manual / Borg-Warner 65, 3- speed automatic

Gear	manual Ratio	manual mph / 1000rpm	Automatic Ratio	Automatic mph / 1000rpm
Top	0.833	27.55	1,000	22.95
4th	1.000	22.95	—	—
3rd	1.396	16.44	—	—
2nd	2.807	8.18	1.46	15.72
1st	3.300	6.95	2.40	9.56

Final drive gear	Hypoid bevel
Ratio	3.31 3.31
SUSPENSION	
Front—location	Double wishbones
springs	Coil
dampers	Telescopic
anti-roll bar	Yes
Rear—location	Lower wishbones and fixed length drive shafts
springs	Coil
dampers	Telescopic
anti-roll bar	No
STEERING	
Type	Rack and pinion
Power assistance	Yes
Wheel diameter	15¾ in.
Turn lock to lock	3.33
BRAKES	
Circuits	Dual, split front / rear
Front	11.18 in. dia. disc
Rear	10.38 in. dia. disc
Servo	Vacuum
Handbrake	Centre, on rear discs
WHEELS	
Type	Pressed steel disc
Rim Width	6 in.
Tyres — make	Dunlop SP Sport
type	radial, tubeless
size	ER70 VR 15 in.
pressures	F 27 R 26 psi (normal driving)
EQUIPMENT	
Battery	12 V 66 Ah
Alternator	45A
Headlamps	Halogen, 4 lamp system 220/110 W
Reversing lamp	Standard
Hazard warning	Standard
Electric fuses	12
Screen wipers	Two-speed intermittent
Screen washer	Electric
Interior heater	Air blending
Air conditioning	Extra
Interior trim	Leather seats, nylon cloth headlining
Floor covering	Carpet
Jack	Screw scissor type
Jacking points	Two each side under sills
Windscreen	Laminated
Underbody protection	Flintkote type; wax injection
DIMENSIONS	
Wheelbase	
Track: front	58.3 in. (1480 mm)
rear	58.88 in. (1495 mm)
Overall length	195.2 in. (4959 mm)
width	69.7 in. (1770 mm)
height	54.0 in. (1377 mm)
Ground clearance	7.00 in. (178 mm)
Kerb weight	4033 lb (1830 kg)
Weight distribution at kerb weight	F/R 53/47

JAGUAR OF COVENTRY

MOTOR SPORT visits Browns Lane to try the latest models from Jaguar-Rover-Triumph Ltd.

IN VIEW of the release of the Series III models from Jaguar and Daimler a few days ago, after the Press had sampled them in Torquay, the Company's first model-launch since it was established as BL's specialist division, it seemed opportune for MOTOR SPORT to pay another visit to the Jaguar factory at Browns Lane, Allesley, near Coventry to discuss these cars, how they were planned, and examine some of their exclusive features. Andrew Whyte, latterly Product Manager of Jaguar-Daimler Marketing, was happy to lay this on, in his usual efficient manner, for he is a 100% Jaguar enthusiast and historian who was apprenticed to the Company in Sir William Lyons' time. He had ready for us two of the impressive revised Jaguars, a fuel-injection five-speed 4.2-litre with the long-lived twin-cam six-cylinder engine, and a 5.3-litre V12, also with fuel injection, and automatic transmission. C.R. writes about both these fine motor-cars later in this article.

I also drove briefly both these Jaguars and would remark that the Six is so good, and so incredibly flexible with its injection power unit, that anyone unable to afford twice its number of cylinders should nevertheless be quite happy, even though the vee-twelve-cylinder-engined car is the Ultimate Jaguar. If I remark on the outstandingly silent running of this 12-cylinder model and its companion-in-luxury, the Daimler "Double-Six", I may well be told by exacting readers that there is no such thing as a completely-silent motor-car. So may I just say that no automobile ranks higher in the No-Noise stakes than this splendid Jaguar, for whose mechanical near-perfection and up-dating Harry Mundy, the Engineering Director, Powertrains and Transmissions, is responsible, while Jim Randle, Director of Vehicle Engineering, supervised the coachwork transformation. It is an astonishing and very satisfying experience to come from a quiet car, by other maker's standards, to this big-engined Jaguar (or Daimler) and to waft away in unimagined hush, a hush which persists right through the speed-range. Moreover, this V12 from Coventry is the sort of car in which you find you are doing 100 m.p.h. when the intention was to cruise at 70 – perhaps this is why a Cruise-Control is now an

The Editor and his Deputy with Jaguar's Andrew Whyte and a pair of new Series III XJ's – 4.2 left, 5.3 right – outside the home of one of the World's finest marques.

option – closeted in luxury and comfort, yet with entirely effortless acceleration even from, say 80 to 100 m.p.h. and beyond. At the other end of the scale, the latest fuel injection 4.2-litre Jaguar will pull away in 5th-speed from a crawl at which its tachometer-needle was indicating a mere 200 or so r.p.m., as a demonstration of what has been done for this now-aged, but unbowed, long-stroke in-line "six".

At lunch Andrew had brought together to meet us some of the most important people in the newly-established British Leyland section. On my right sat Bob Knight, Managing Director of Jaguar, and among those present were Harry Mundy and Jim Randle, previously mentioned, Mike Beasley, the Manufacturing Director, and Peter Taylor, a senior development engineer. Conversation ranged over production methods here, in Germany, and in America, production targets, paint problems, the steady emergence of the Hassan/Mundy V12 engine in its latest form, and the up-dating of the former Jaguar models. These, and the styling alterations, which include a higher roof-line, new wrap-round bumpers, increased glass area, and improved interiors with the accent on even more luxury and quietness (from better sound-damping), will be dealt with below. The Series III cars represent an investment of more than £7,000,000 and it is because I regard them as such fine cars that this feature about them seemed to be desirable, at a time when the Buy British campaign may – and certainly should – be gaining ground. Of them Bob Knight has said: "When we designed the new Jaguars and Daimlers we faced the difficult task of improving upon products which have already earned a World famous reputation for their performance and refinement. The new range is intended to be evolutionary, in concept. By applying the latest techniques in advanced engineering and many more luxury features I am sure we have added an extra touch of excellence." Having been out in two of these latest Jaguars I was fully prepared to drink to that, before going into the factories to look at some of those special items that go to make a Jaguar (or Daimler) such a very special kind of car.

* * *

I had been over the Browns Lane Jaguar plant previously, as had C.R., so we confined ourselves on this occasion to looking at the trim and upholstery shops, with just a quick visit to the long assembly lines, from where, last year, more than 26,500 cars emerged, 1,700 more than in 1977, with nearly 5,000 of these going to American buyers, an increase in USA sales, over those in the previous year, of 10%.

The size of the trim and upholstery departments at Browns Lane is visual evidence that Jaguar has no intention of giving up the traditional leather seats and veneer finish in its cars. Whereas some stylists try to blend in the tree-wood (or a horrid synthetic substitute) with the plastic, or slyly add a strip or two of wood here and there in an otherwise undistinguished decor, Jaguars and Daimlers continue to have fully-veneered, beautifully-finished interiors and leather upholstery; and it looks as if they always will. Indeed, leather is a no-cost option on all models except the 3.4 XJ6, for which it is an extra, and it is standard on the 4.2 and Daimler Vanden Plas models. In fact, some 90% of customers ask for leather. This calls for a large shop in which to produce this adjunct to a pleasing-to-own status motor-car. The leather is supplied by Connolly's, who deliver something like 1,500 hides a week, of which about 2½ hides go into the upholstering of each car. The use of templates when cutting out the hides into the required patterns obviates waste and whereas plain leather is used for the Daimler and Vanden

Bob Knight, Managing Director.

Mike Beasley, Director of Manufacturing.

Plas bodies, that for the rest of the range is embossed, which enables a slightly less-perfect hide to be made use of. The suitably-shaped upholstery is then sewn to the foam of the upholstery. In the preparation area the hides are got ready for dispatch to the equally large trim shop, where ordinary tough Singer sewing-machines are much in evidence. 600 operatives are employed here, including 31 girl-machinists (it is traditional to have ladies for this work) using a total of 60 sewing-machines, mostly these quite aged Singers.

The wood-working shops, known as The Sawmill, are another very special feature of the Jaguar plant. Dashboards really are dash*boards,* shaped from birch "through-and-through" plywood, with the apertures for the instruments, etc. formed ingeniously with electric rotor machines as the operatives nimbly negotiate templates round fixed pegs on the machines' flat-beds. The highly-polished, Californian-grown, figured-walnut veneers, brought from France, are matched from the centre outwards on a given panel. Slightly darker veneers are favoured for the Daimler Vanden Plas bodies, which are soon to adopt high-quality burr-walnut as used on the big Daimler limousines. All the veneer selection and matching is done by one pair of experienced eyes, those of Bill Burke, who has been with the company since the war years and remembers working on Swallow sidecars. Blanket presses are used in the veneering process, some of which are nearly 30 years old. The pressing occupies only a few minutes for most jobs, up to quarter-of-an-hour as a maximum. Man-made veneers have been looked at, but we got the impression from the scathing comments of Sawmill Foreman Frank Burrell that they are not going to be seen on Jaguars, at all events not in the immediate future. Before veneering the plywood panels are flatted off and their edges and apertures sealed. After the veneer has been applied it too is flatted off and a sealer coat applied. Then follow two coats of polyester and two grades of flatting before the surface is polished twice on an automatic machine to give a burnished wood finish. The beautiful burr-walnut for the limousines is polished with mops by hand. Jaguars sought for many years to overcome the effects of the sun on woodwork and this was achieved by the foregoing type of finish, introduced with the Series II XJs. Now, if there is any effect at all it is for the woodwork to darken rather than be bleached or the varnish to crack and peel as before.

The assembly lines at "The Jaguar" are another impressive sight. We have described them previously; two floor-level tracks now operate for the saloons and XJ-S coupes, these running the length of the vast, well-lit assembly area, then turning to go in the opposite direction as the cars take shape. Jaguar engines were traditionally fitted by lowering the body onto the ready-assembled engine/front suspension/subframe assembly, a system started with the first unit-construction car, the 2.4, in 1956. This was all very well with the Solex-fed 2.4 engines, but with the later SU-fed Mk. 1 3.4s and 3.4 and 3.8 Mk. 2s, the carburetters had to be removed from the fully-assembled engines to allow clearance between the longitudinal frame members. The fitting of the engines from below continued with the introduction of the XJ6, but was not possible with the broad V12, which had to be lowered in from above, the production line having to be modified for this purpose. Only very recently has this simpler process been adopted for the six-cylinder cars.

Output of Jaguars and Daimlers is presently running at approximately 600 to 640 saloons and 75 to 100 XJ-S a week, a total of 30,000 a year. The expectation is that next year output will be up to 32,000 to 35,000 cars. In 1978 13,500 cars were registered in the UK. The bodies come to Allesley from a new £15,500,000 paint plant at the Pressed Steel Fisher factory at Castle Bromwich, which is housed on two working levels covering an area of 500,000 square feet. The improvement in the Jaguar paint process with the introduction of the Series III XJs is most significant, for paint finish has always been the main point of complaint from Jaguar customers. I will leave it to C.R. to describe briefly the new process and the new range of colours, both shared by the XJ-S, in his description of the new cars.

Part of the self-contained trim shop with its battery of faithful Singers.

One significant point on which to conclude my brief survey of all being well at Jaguars. Now that the great vee-twelve-cylinder engine is fully established and petrol injection has improved its fuel thirst from the original 9 to 11 m.p.g. apparently to something like 15 m.p.g. (full information when MOTOR SPORT has road-tested the latest model), just over 1/5th of the saloons are supplied with this bigger engine. I will now hand over to C.R. for an impression of the engine-assembly plant at Radford, his description of the Series III modifications and his report of driving two of the new Jaguars. – W.B.

* * *

BY THE time W.B. and I had tried the Series III cars, lunched with the directors in the oak-panelled dining room and toured the Browns Lane Factory, the workers at the Radford engine plant had dispersed, for the normal two-shift working had been reduced to one while production of the new models was wound up to full swing. Thus Andrew Whyte and I had a lonely visit. A tour of silent, idle machinery is much less informative than the active process of components taking shape, so much of the information on vee-twelve production which follows later has been gleaned from Jaguar's own sources.

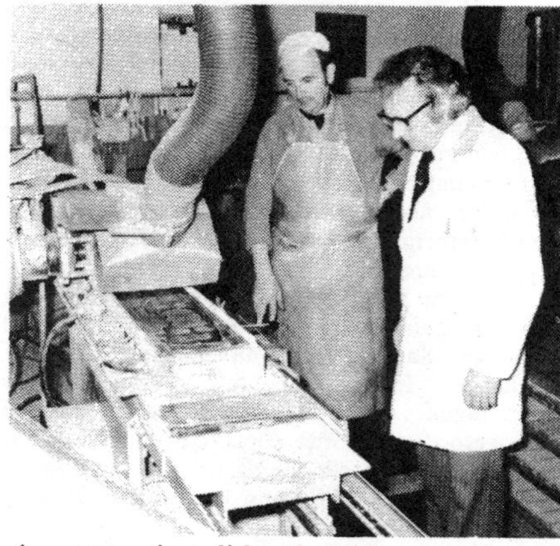
An automatic polisher burnishes a walnut XJ facia, watched by Bert Allbrown, left, and Sawmill foreman Frank Burrell.

The former Daimler factory was liberally plastered by German bombs and re-constructed post-war. When W.B. described a visit to it in MOTOR SPORT, July 1966, its vast area housed an extraordinary production mixture of Daimler Majestic Majors, their 4½-litre V8 engines, 2½-litre V8 engines for the 250 saloon, Jaguar in-line six engines and Daimler Fleetline 'buses. Now Daimlers are either re-badged and re-grilled Jaguars assembled on the mixed production line at Browns Lane, or limousines crafted at the Vanden Plas factory in North London, while the Fleetlines bear Leyland badges and emanate from Lancashire. But the sense of history has not been lost, for the factory showroom housed a privately-owned SP250, the Queen Mother's old limousine, recently replaced on the Royal fleet by a new example, and a 1911 23 h.p., six-cylinder, sleeve-valve Daimler. I was also pleased to see there Jaguar's immaculate, black XK 120 roadster, temporarily removed from the Browns Lane entrance hall, which a recent dealer convention had caused to be temporarily stripped of its inspiring Jaguar collection.

Besides engine production, the Radford factory is used for suspension subframe assembly and a liberal littering of MIRA-smashed, orange painted Jaguars reveals that part of the Engineering Department is housed there too.

Production of the long-lived Jaguar XK engine has changed little since W.B. described it briefly in 1966. The same battery of individual machines survives, a contrast with the huge transfer machines installed in 1971 for vee-twelve production. But crankshaft journals and camshaft profiles are now machine- instead of hand-lapped and a Fel-Electric Crack Detector has been installed. Crankshafts are Tuftrided, con-rods balanced end to end and pistons and con-rods installed as matched, balanced sets. Each engine is bench-tested with its gearbox on a row of ageing Heenan and Froude DPX3 dynamometers, firstly being started and run unladen to check for oil leaks etc., and then run for 45 minutes at 1,200 r.p.m and 25 lb. load, a rather less rigorous test process than the six hour cycle employed in 1966.

The fully-automated vee-twelve production line cost £3,000,000 when installed and would doubtless cost several times that today. W.B. described much of its operation in detail when writing about the new E-type V12 in MOTOR SPORT, April, 1971. The machining and assembling procedures have been practically unchanged since then, save that the operatives on the moving carousel assembly line now bolt on Lucas-Bosch D-Jetronic fuel injection instead of quadruple Zenith-Stromberg carburetters. New manifolding is cast and machined to suit.

The major items of equipment are three Archdale transfer machines with 57 stations between them. Equipment for machining the cylinder heads includes a 42-station Huller transfer machine. Crankshafts, connecting rods, bearing caps and other machined components are all done at Radford. There being no foundry here, the aluminium cylinder blocks and heads are machined from castings supplied by sources such as Birmid. Five special-purpose Cincinnati machines carry out initial milling operations on the block, including the sump, head, top, bearing cap seat and end faces. The three Archdale machines then carry out the main drilling, reaming, tapping, milling and boring operations. A Weatherly horizontal broacher is used for sump face and bearing cap locations and Desoutter and Ingersoll-Rand stud insertion and nut runners fit bearing caps. Other equipment used includes a GFM twin-headed miller, a Kearney and Trecker transfer machine, Landis grinders, a Boncham and Turner fine borer and Avery balancing equipment for the reciprocating parts.

For a more detailed description of the fine-tolerance production of this magnificent engine I refer you to W.B.'s aforementioned article in 1971. Minor changes have been made to the dynamometer run-up procedure since then: after being started without load they are run with 35 lb load at 2,000 r.p.m. for 30 minutes, followed by 80 lb load at 2,500 r.p.m. for 15 min.

All this machining takes place where the Majestic Major was at one time assembled, and Sp250 glassfibre bodies produced. After completion and testing the engines are stored in steel pallets in the lofty adjacent shop in which the Fleetline 'buses were built. An asbestos partition between the two walls is a reminder of the flying welding sparks from the construction of armoured cars.

Leyland lorries carry the six- and twelve-cylinder engines from Radford, where Jack (no relation to Jim) Randle is Plant Director, the

One of the huge automatic transfer machines for vee-twelve engine production.

three miles to Allesley, the domain of Wal Turner. Quality control at both plants is under the Directorship of David Fielden.

Almost round the corner from the Radford plant is the Jaguar-Daimler Service Department, which is expanding under the direction of Neville Neal. This is interesting, because it denotes a change of policy within BL, part of the recognition that the Jaguar *marque* name must be kept strong and divorced from the rest of the BL range. In earlier Leyland days there was every sign that Jaguar would lose its individual character, of which the reduced facilities of the Service Department after its move from Browns Lane was symptomatic. No longer did Jaguar have its own service engineers providing direct link between dealers, and thus customers, and there is no doubt in the minds of some long-serving Jaguar employees that this disruption of liaison did the marque no good at all. Now the situation has been reversed: the company has its very own specialised service engineers once more instead of sharing those of the rest of the Group. The Service Department has a direct function too, servicing customers' cars – mostly local – and dealing with difficult problems on brand-new or warranted cars referred to them from dealers by the service engineers.

Of course, the aim of Jaguar engineers is to cut down service problems and warranty work as much as possible and on that note of optimism I will move on to Jaguar's hope for the immediate future, the new Series III XJ6 and XJ12.

The new bumpers and grille are instant recognition factors on the Series III. This example is the XJ6 4.2 manual.

THE NEW SERIES III JAGUARS AND DAIMLERS

ALTHOUGH it is eleven years since the XJ6 was introduced, the Jaguar/Daimler model range remains unsurpassed in its combination of ride comfort, suppression of road, suspension and wind noise and roadholding. How could Bob Knight and his team of engineers improve on this near perfection? The answer has been to make the coachwork itself more refined whilst leaving the chassis practically unchanged save for fitting stiffer steering rack mounts to improve response when turning into a corner. In the case of the 4.2-litre car, the adoption of fuel injection has put back the performance which increased weight and restrictive emission regulations had sapped away over the years and a five speed manual gearbox is available for those customers who want a more sporting alternative to the Borg-Warner Model 65 automatic transmission on the XJ6 3.4 and XJ6 4.2.

At a quick glance the new Series III is not all that different to the Series II, the most obvious change being the wrap-around bumper bars clad with black injection mouldings. Closer examination reveals that the profile of the "glass-house" area above the waistline is completely new. At the rear the roof has been heightened to give passengers more headroom, so important in a car which is often chauffeur-driven. As a result the roof is less rounded and more angular when viewed from the side and the roof-line sharper. The backlight is flatter, yet generous parcel shelf space remains. The screen pillars are raked three inches further forwards so that the higher roofline actually looks sleeker than the original shape. Increased tumble-home of the side-windows has been created by reducing the width of the roof. Deeper windows all round have considerably increased the glass area. The front quarterlight pillars have been removed. Standard equipment laminated windscreens are fitted by direct glazing, which has an added bonus of increasing bodyshell torsional strength. Tinted

The latest version of the Jaguar XK engine has Lucas-Bosch fuel injection and produces 205 b.h.p.

This photograph shows the revised silhouettes, flush-fitting door handles, new wheels and the omission of the front quarterlight pillar.

SERIES III JAGUARS –
continued from page 470
glass is standard on all models except the 3.4, on which it is an optional extra.

Another obvious external change on the Jaguars is a new radiator grille with vertical bars and a centre rib, shared by both the XJ6 and XJ12; the latter has lost its central "V" badge and is distinguished merely by having a gold on bronze Jaguar head mascot at the top of the grille instead of the XJ6's gold on black. Daimler grilles are unchanged. Door handles are lift-up and flush fitting, revised rear lamp clusters incorporate reversing lights and a flatter, wider number plate light housing incorporates the boot lift latch. Large, door-mounted, twin exterior mirrors, available with either electric or manual remote control, are a standard fitment. I thought the new black and stainless steel wheel trims of the XJ6 we tested as attractive as the optional alloy wheels fitted to "our" XJ12.

Those then are the obvious differences. General equipment has been considerably improved too, but more of that later. First let me deal with the revised mechanical details of the XJ6 4.2, the most significant change in this new range. Over lunch Bob Knight and Harry Mundy made no bones about their concern and awareness during the run of the Series II 4.2 that its performance had begun to be sadly lacking. Indeed at one stage its power output had dropped below 170 b.h.p., within a handful of b.h.p. of the 3,442 c.c. engine, which continues unchanged in Series III form, producing 162 b.h.p. DIN at 5,250 r.p.m. Mundy's modifications have given the 4,235 c.c. engine an enormous boost to 205 b.h.p. DIN at 5,000 r.p.m. to set this model well apart from the bottom of the range car.

Ironically, after the emasculation of Jaguar power by emission controls, this big improvement results from the use of an engine developed to comply with US emission regulations and which has been fitted to Series II US cars for a little while. Its main feature is the fitment of Lucas-Bosch electronic fuel injection, a joint Jaguar and Lucas development of the K-Jetronic system. A single high pressure pump fed from the twin fuel tanks maintains a constant 36 p.s.i. pressure to the injectors relative to the intake manifold pressure. The heart of the system is an electronic control unit connected to sensors on the engine. Fuel injection pulses are triggered from the Lucas Opus electronic ignition system and are produced every revolution instead of every cycle as on the D-Jetronic system used on the V12 engine. A separate automatic cold start circuit is incorporated. A special economy feature on 4.2-litre cars with automatic gearboxes is an overrun fuel cut-off which stops the fuel supply to the engine when the throttle is released until revs drop to 1,200 r.p.m.

Other improvements to performance have been brought about by raising the compression ratio to 8.7:1 and fitting larger 1.875 in. inlet valves. Induction noise has been reduced by removing the hot air flap in the air cleaner which was necessary for carburettor fuel control.

GM 400 automatic transmission is standard equipment on all XJ12 5.3-litre Series III cars

and the majority of 3.4-litre and 4.2-litre cars will be fitted with Borg-Warner Model 65 automatic transmission, compulsorily on the Daimler Vanden Plas 4.2. For the minority of Jaguar-Daimler owners who prefer manual transmission (only 10 per cent at the last count), Harry Mundy and his men have developed a version of the Rover 3500 five-speed gearbox to replace the four-speed overdrive unit previously fitted. Incidentally, an announcement last yea that the five-speed 'box had been made availabl on the Series II was premature. Modification required to make the 'box stand up to the extr demands of the Jaguar power included bigge Timken bearings and a stiffer tailshaft. Jagua designed their own, shorter shift gearchang mechanism too, which is machined at Radford shipped to Pengam in Wales for assembly on t the gearbox and then shipped back to Radford t be mated to the power unit. Mundy insisted o having a lift-up reverse protection instead o knock-over – anybody who has felt the ease witl which well-used earlier Jaguar gearboxes particularly pre-synchro units, could be slippe into reverse inadvertently will realise why. Botl manual and automatic 4.2s will have a 3.31:1 final drive ratio against the low 3.54:1 fitted in later series II cars and the latest 3.4.

With this new gearbox comes a new hydrauli clutch release system, a hydrostatic arrangemen in which the lines are always full and unde slight pressure, so that the thrust race is alway lightly in touch. The old carbon thrust washe has bitten the dust at last.

Coincidentally with the arrival of this nev model, Jaguar have adopted a totally revisec paint finishing operation, carried out in a nev paint shop alongside the Pressed Steel body build plant at Castle Bromwich. The new process involves the use of ICI thermoplastic acrylic paint, of which four coats are applied automatically after much more thorough preparation than previously. This TPA is baked at much higher temperatures than the old paint finish and is claimed to give a better, more even, harder yet flexible finish. The bodies are final-finished before delivery to Browns Lane for assembly instead of receiving their final spray coat after assembly and road testing. Each car continues to be individually road tested, however. All Series III inner panels, the longitudinal chassis members and the doors are wax injected for corrosion protection.

With this new paint process comes a much wider choice of colours. Nine high-gloss colours, plus black, will be progressively introduced, with the identifying point of a single coach-line on the 4.2, a double coach line on the 5.3 models. Five metallic colours will be offered: apart from silver, metallic colours were dropped from the XJ range early in its life because of problems with them. XJ-S coupes will also utilise the new paint process.

Driving the Series III Jaguars

ANDREW WHYTE first of all proffered the keys

The detail tail-light and trim differences are shown in this comparison of the new 4.2 on the left and its immediate predecessor.

of a Jaguar XJ12 5.3 (this is the full nomenclature on the Series III boot-lid). It is almost impossible to be subjective about a car which was so very nearly perfect when we tested it in earlier guise. Sensitive instruments might detect that this latest V12 is quieter than the previous model, for all the Series III cars have improved sound deadening, but when the quietest car in the World is made quieter still, the human ear can barely tell the difference. The only obvious wind noise came from the twin, electrically-adjustable door mirrors.

The immediate impression was of a lighter, airier, less claustrophobic cockpit, with much improved all-round visibility and mirror vision. The somewhat reflective veneered facia and instruments are unchanged. The so-comfortable front-seats in all the models have lumbar support adjustment over a range of 1½ in. Both W.B. and I felt the seating position to be a little low, especially now that the increased headroom allows more latitude. Electrically controlled drivers' seats, incorporating height adjustment, are available as an extra. Both front seats on the Vanden Plas models have electric control as standard.

At last Jaguar have fitted self-parking wipers to their saloons, the column stalk also offering intermittent wiping.

With a car so smooth and quiet as this 287 b.h.p. V12 Jaguar, even the most cautious of drivers can slip unwittingly to very considerably over the speed limit. Econocruise speed control, fitted to the test 5.3, is a very welcome optional extra on this model and the 4.2 automatic. It is switched on by a rocker switch on the facia and set at the required speed by a very awkward-to-operate button in the left hand steering column stalk. The setting can be over-ridden by throttle or brakes and the set speed resumed by pressing the central rocker switch. It works smoothly and efficiently, though not so conveniently as the single-stalk Porsche cruise control.

This 5.3 pulled an indicated 100 m.p.h. at 4,000 r.p.m. in virtual silence and its astonishing acceleration on full kickdown was accompanied by a modest volume whirring from under-bonnet. The steering was creamily effortless – many would say overlight – yet positive enough and responsive. It employs the seven-tooth rack as before on the later V12s and XJ-Ss and this higher steering ratio has now been adopted for the six-cylinder Series III cars.

There is not much more I can add to what W.B. has said about our brief drive in the XJ12 5.3, a sublime way to travel which we have experienced and written about before. The 5-speed manual XJ6 4.2 was more interesting, almost a return to a real Jaguar sports saloon, only slightly less hushed than the V12 and with quite vivid performance compared to the earlier 4.2.

The extra power, improved throttle response and better control through the well-spaced gear ratios gave the impression through both performance and handling that this Series III had lost half a ton in weight. W.B. has already described the extraordinary flexibility of this long-stoke "six", which matched the Rover Turbo tested elsewhere in this issue in its refusal to stall in fifth gear, though there was a detonation when the throttle was floored.

The gearchange of this low-mileage car was still loosening off, but its positive, shorter movements were preferable to those of the Rover 3500 and my only real criticism was of the closeness of the central console, which my elbow caught during quick cog-swaps. Clutch action and sensitivity was a great improvement over the heavy, dead-feeling mechanism of old.

The more I drove this middle-sized manual Jaguar the more I liked its excellent performance, roadholding which remains second to none in its field, first class brakes and the lightness of its handling. All this accompanied by unbeatable comfort of ride, and whispering progress. No doubt about it, these latest improvements have converted what had become a stodgy fine car into a lively fine car.

A few other details worth mentioning is that electric windows, central locking, interior light delay, a timed heated rear window as on Mercedes, an electric aerial delay to stop it whirring up and down every time the starter key is turned, halogen headlights and radio/stereo cassette are all included in the package. Extras available include headlamp wash/wipe, an electrically operated sun-roof and air-conditioning (standard in the Vanden Plas). – C.R.

Jaguar-Daimler Series III Price Range

Jaguar XJ6/XJ12	
3.4 auto or manual:	£11,189
4.2 auto or manual:	£12,326
5.3 auto:	£15,015
Daimler Sovereign/Vanden Plas	
4.2 auto or manual:	£12,983
Double-Six auto:	£15,689
Vanden Plas 4.2 auto:	£17,208
Vanden Plas 5.3 auto:	£20,277
Examples of extras:	
Air conditioning (standard on Vanden Plas):	£918
Electrically controlled driver's seat (both front seats electrically controlled as standard on Vanden Plas):	£150
Electric sun-roof:	£450
Cruise control:	£250

These prices were obtained as we went to Press, so no comment is included in the text.

Still on the Jaguar theme, this XJ12 saloon appeared in a supporting race to the South African GP.

The very last E-type coming off the line in 1971.

Inflation

When Jaguar Cars stopped making the E-type they ended the run with a batch of fifty Series III V12 cars, the last of which the firm kept for their own publicity purposes. The forty-ninth of the series was sold to a private Jaguar enthusiast and was painted green, while all the rest of the batch were painted black. Each car had a small brass plate on the instrument panel, giving the car details and inscribed with Sir William Lyons' signature. This was in 1975 and the cars were sold at the normal V12 E-type price of £3,743 for the manual version and £3,937 for the automatic.

A motor dealer in Surrey bought two of these black V12 E-types, numbers 2823 and 2845, presumably at a trader's discount price of 17% under the list price. He salted them away, unregistered, unused and brand new, and kept them in an air-conditioned showroom. He is now offering them for sale at around £25,000 each!

Having driven the very last E-type V12, Registered Number HDU555N when it left the production line I wept no tears at the end of the E-type, for by 1975 it was terribly dated. The V12 engine was fabulous, but the rest of the car could not cope with it. If you used the potential of the 5.3-litre V12 engine you soon ran out of brakes, the road-holding was sadly lacking, the steering was only just adequate and the whole car was very old-fashioned in its manner of going, though still a super car to look at. Tempering one's driving style to suit the car, as regards a feeling of personal safety and well-being, I found that one was using the potential of a good 4.2-litre 6-cylinder E-type (like the one I was using at the time). When I returned the V12 roadster to the factory I said I thought the engine was terrific but it needed a new car around it. The Jaguar men just smiled and said nothing, for they knew they had the XJ-S coupe about to go into production. As we now know the XJ-S can use the potential of the 5.3-litre V12 Jaguar engine in all respects, and after you've driven one you know it is a good car.

The Series III V12 E-type was really only a stop-gap model used to get the V12 engine into production before it went into volume production for the XJ saloon and the XJ-S coupe. In effect the V12 E-type was a test-bed for the production engineer at Jaguar cars, otherwise the E-type would have ended with the Series II 4.2-litre 6-cylinder. It had been in production for over ten years before it became dated which is the sign of a successful car.

The Surrey motor dealers who want £25,000 for a car they bought for under £4,000 four years ago are obviously not looking for a motorist as a customer. Anyone who knows about motoring would buy an XJ-S coupe for a lot less money and have enough left over to pay for using it for 50,000 miles or more. You could even buy a Porsche 928 and have money left over.

Are "collectors" lunatics, clever, dishonest, good business men, endowed with more money than brains, or perhaps they don't really exist, except in the imagination of the motor dealers? If these two nondescript cars from the final batch of 50 E-types are worth £25,000 each to the dealers, then one wonders what the Trade would ask if they could get their hands on HDU555N, the very last production E-type Jaguar – D.S.J.

JAGUAR SERIES 3

More interior space, enhanced comfort, revised looks and fuel injection are features of the new models, International Editor GORDON WILKINS reports from Britain.

A new roof, with increased rake for screen pillars, more "tumble home" for side windows, more rear headroom and a flatter, deeper rear window are the main structural changes in the up-dated Series III Jaguars. With flush door handles, simpler grille, deeper black-faced bumpers, revised wheel trims and a tidier rear end with bolder tail lamps, the cars present a much-more-modern appearance.

The 4,2-litre "six" now has Lucas-Bosch electronic ignition producing an extra 22 kW, and a strengthened version of the Rover 3500 five-speed manual gearbox with a mass about 13 kg less than the four-speed with overdrive. As before, the 4,2 can be had with the Borg-Warner 65 automatic transmission, while the V-12 has the GM 400 as standard. Cheapest model in the range, the 3,4 "six", has the new body improvements and a five-speed or automatic transmission, but retains two SU carburetters.

A redesigned interior, many new items of equipment and still better sound insulation enhance the attraction of the new models. A new moulded head lining with recessed visors helps to extend the head room which is now generous by any standards and a sun roof with manual or electric operation is a new option. Electric control is also available for seat height and door mirror adjustment.

Steering column adjustment is included, but the greatest contribution to reducing driver fatigue is a simple variation in lumbar support which can be adjusted in or out over a range of about 3,8 cm by turning a handwheel. There is a new leather-covered steering wheel, instruments are easier to read, and switches are identified by international symbols instead of words.

Automatic-transmission 5,3 and 4,2 cars can now have a cruise control, with "On/Off" and "Resume" switches on the console and speed selector button on a steering column stalk. Its memory is erased when the engine is switched off. The large outer headlights, now with halogen lamps, can be fitted with electric wash/wipe, sharing a 7-litre reservoir with the windscreen washer, and operating only when the headlights are in use. The windscreen wipers offer both interval and flick-wipe facilities. Transistor switching gives a 10-15 second delay to keep the interior lights on and enable the driver to locate other switches after entering the car, and there is a time switch to cut out the heated rear window automatically 15 minutes after switching on.

All Series III models have an electrically-erected and retracted radio aerial and there is a time switch with 10-second delay which prevents the motor operating while the engine is being started. Radio and stereo cassette players and four speakers are included in the standard equipment.

The Series 3 Jaguar (left) has a revised grille and headlight washers. The instrument panel and console (top) and fitted toolcase in the trunk (above, left) are new features, and the tail lights (above, right) are re-grouped.

On the instrument panel, there are now warning lights for low coolant level, rear fog lights and bulb failure in brake lights or the side lights, which are now incorporated in the headlights. These are in addition to the usual monitors for main beam, hazard warning, handbrake, brake fluid failure, alternator, oil pressure and seat belts. In the trunk is a new moulded case with carrying handle, containing a set of plated tools, spare lamps and fuses.

Equivalent Daimler models continue with all the improvements, and the top Daimler Vanden Plas, with vinyl roof, as 4,2 or Double Six, incorporates most of the options as standard equipment, including air conditioning.

It is no longer necessary for a car to be very low-built to provide a high performance, and the new Jaguars give the impression of being easier to enter and lighter and more spacious inside. The first impression is helped by the base of the screen pillars being moved forward, and the roof panel being narrower. The second impression is helped by the larger glass areas and by the elimination of separate quarter windows in the front doors.

Jaguars have often been deficient in lumbar support, but new seats offering adjustable contours for the backrests, in addition to the other possibilities, put them right in the front rank.

The improved sound insulation puts both six and V-12 models in the top class for silence and refinement. Wind noise remains low even at 160 km/h and over, and the ride is first-class, with only a slight firmness at low speeds as evidence that there are any bumps in the road at all. Both "six" and "twelve" were showing about 4 000 r/min at an indicated 160 km/h, and with the right conditions would accelerate briskly to an indicated 190.

Jaguar's power steering is light but gives adequate feel, and on the "six" the five-speed gearbox, with its effective synchromesh, seems a more modern solution than four speeds with overdrive. With automatic, the "six" — producing its peak torque at 500 r/min less — seemed to be changing up and down somewhat less on narrow, winding, hilly roads, but overall performance of the "twelve" is considerably higher. The electronic cruise control works very smoothly, and pressing the "Resume" button to regain speed after braking gives a very relaxed style of driving for motorways.

New colours are among the attractions, in the latest acrylic resins, applied in Britain's most modern paint shop, adjacent to the Pressed-Steel-Fisher plant where the bodies are built. It cost R27 million, plus another R7,8 million for associated services. The treatment includes a special adhesion-promoting coat on top of the two coats of primer surfacer, before the four colour coats are applied automatically. The complete body shell is then spirit-sanded to produce a blemish-free surface before the final re-flow operation in the oven. Further protection is provided by underseal and by wax injection into critical parts of the shell. The air filtration system for the paint booths changes air at the rate of a 100 000 cubic metres per minute.

As part of the new quality control system, paint film thickness is monitored continuously, and bodies selected at random are given a ten-day salt spray test. ●

Star Road Test

DAIMLER SOVEREIGN SIII 4·2

More grace, pace and space than ever in the subtly restyled Series III, and improved economy into the bargain. Virtues it has in abundance, but some traditional shortcomings remain

IN MARCH this year, following a very successful 1978 (26,500 vehicles produced, an increase of 1,700 over 1977), and two months before the 10th anniversary of the introduction of the XJ6, the new Series III versions of this Jaguar/Daimler range were announced. Jaguar Rover Triumph's aims with the Series III were to improve on an already excellent product, which is without question the most refined motor car ever built, to broaden its appeal with a number of important modifications, and to modernise its appearance.

The Series III is most certainly an example of evolution rather than radical change, though it may surprise many people to learn that the only body panels carried over from the Series II are the floorpan, boot and bonnet. The main changes are to the roofline, which has been raised to improve interior headroom, and to the windscreen pillars which have been brought forward at their lower ends, resulting in a more steeply raked windscreen. The side windows are deeper than before and there are now black wrap-around bumpers front and rear, new rear light clusters, and neatly recessed door handles. The external dimensions of the new car are unchanged.

Jaguar versions have a new grille, which recalls company tradition with its vertical bars, while the Daimler name and fluted grille live on. The Daimler Sovereign 4.2, which is the subject of this test, costs £14,516, slotting in between the mechanically similar Jaguar 4.2 (£13,781) and the 5.3 XJ (£16,788). Daimler versions of the XJ saloons are better equipped and offer an added touch of luxury.

Mechanically, there is one important change to the 4.2-litre engine, the adoption of Lucas/Bosch fuel injection, considerably modified in detail from the system which has been used for the past four years on the V12 engine.

The use of injection and modifications to the induction system and cylinder head have boosted maximum power by 25 bhp, though peak torque is unchanged. The effect of this extra efficiency has been to improve both performance and economy, by a considerable margin.

The most significant option (at no extra cost) is the excellent five-speed Rover gearbox, now available on all six-cylinder models (V12s continue to be automatic), but our test car was fitted with the Borg-Warner 65 three-speed auto, which has recently undergone several important modifications to improve change quality and response.

The recent rise in the value of the pound cannot make Jaguar's management too happy, but in this country at least the cars offer supreme value for money to the men (or, more usually, the companies) with the means to run them.

Its main rivals are German: the BMW 733i, at £14,480 (£531 extra in automatic form) and the Mercedes-Benz 350 SE (£16,840). A more expensive alternative is Cadillac's Seville (£18,232), while among the less costly ones are the Opel Senator (£11,260) and the still cheaper Rover 3500 and Ford Granada Ghia.

PERFORMANCE

★★
★★

The origins of the straight-six XK engine date back to 1948, though it has been the subject of frequent refinements. The latest of these, introduced to the US export market last year, and now standard everywhere in the Series II 4,235cc models, is the adoption of Lucas-Bosch electronic fuel injection. This, combined with a raised compression ratio (up from 7.8 to 8.7:1) and enlarged inlet valves, has boosted maximum power from 180 bhp at 4,500 rpm to 205 bhp at 5,000 rpm (DIN figures). Peak torque, 232 lb ft (DIN), is unchanged, but is now developed at 4,500 rpm.

Strict comparisons in performance with the Series II Jaguar 4.2 which we tested in March last year are not valid, because the Daimler in this test was fitted with the higher 3.07:1 final drive ratio, a free option. However, a measure of the improvement in the engine's efficiency is that the acceleration from rest to 70 mph is virtually unchanged, despite the longer-legged gearing, and from then on significantly better.

We recorded a maximum speed (in Germany) of 128.0 mph, compared with 117.5 mph in the older car. From rest to 60 mph took 10.5 sec, using the manual hold on the gearbox; staying automatic adds almost a second to this time, and more than two seconds to the 0-100 mph time of 28.4 sec. Kickdown gets you from 30 to 50 mph in 4.1 sec, and from 50 to 70 mph in 5.7 sec.

These figures are by no means exceptional for the class, for most rivals are substantially lighter. But no other car performs with such a remarkable lack of fuss or undesirable noise. In our view the only car in the world which rivals the 4.2-litre Jaguar/Daimler for refinement is its 5.3-litre V12 big sister.

An automatic circuit, separate from the main injection circuit and governed by a thermostat enriches the fuel/air mixture when the engine is started from cold; this we found to be both efficient and unobtrusive.

One advantage of the adoption of fuel injection has been the elimination of the "hot air flap" which was necessary for carburetter mixture control; at high air flows this tended to perform erratically and its elimination has taken the edge off the slight harshness which exists as maximum engine speed is approached. At 5,000 rpm the engine is still audible, but only just, and the quietness of cruising at 90-100 mph makes the new Daimler a wonderfully relaxing car to drive over long distances.

ECONOMY

★★★

Without question, the XK engine is more efficient with the new injection system than it was on carburetters. No clearer evidence is needed than the fact that our overall consumption of 15.7 mpg, which reflects our usual hard driving, is better than the touring figure of the old car, and is now about par for a class in which many rivals have smaller engines.

Despite great efforts, we were unable to fit our consumption equipment on to the car to take steady-speed figures, but Jaguar supplied us with the results they obtained on our test car, and we trust them. These figures (29.1 mpg at a steady 30 mph, 22.2 mpg at 70 mph) confirm the improvement, and our computed touring figure is 19.3 mpg, which we feel many owners will approach, if not better.

Total tank capacity in the Jaguar range is 20 gallons, carried in two tanks which have to be filled separately, an arrangement which is by no means universally popular. This should give a practical range on four-star petrol of approximately 350 miles.

TRANSMISSION

★★★★

Although still nominally a Borg-Warner 65, the automatic gearbox used in the 4.2-litre Jaguar/Daimler range for a considerable time has been considerably modified for the Series III to the extent that it is sometimes referred to as a 66. Up to now the vast majority of customers have opted for this transmission in preference to the manual gearbox, but it is anticipated that the availability of the five-speed Rover gearbox, with its advantages in choice of ratios and reduced petrol consumption, will attract about 50 per cent of buyers.

In our test car, upward and kickdown changes were both noticeably smoother than in previous XJs we have sampled and in most circumstances the gearbox offers an excellent blend of responsiveness and unobtrusiveness. But two major criticisms remain: the box will not kickdown into first at any speed above about 30 mph, and the change-up point from first to second on full throttle is unnecessarily low (4,400 rpm/45 mph instead of 5,000 rpm/52 mph). Both these factors can make the car seem irritatingly sluggish under certain conditions, such as accelerating out of a roundabout. Under full throttle, the box selects top gear at maximum rpm, equivalent to a road speed of 85 mph.

Most people using automatic gearboxes take the sensible view that manual operation is like having a dog and barking yourself, but those who wish to do so with this box will find it slow to respond to movements of the lever, though the changes that result, particularly downwards, are generally exceptionally smooth.

HANDLING

★★★★

The Sovereign's all-independent suspension has a pedigree stretching back to the E-type and Mark 10, though it has undergone considerable development and refinement since then. At the front, which incorporates anti-dive geometry, there are semi-trailing wishbones, an anti-roll bar and coil springs; at the rear there are lower transverse wishbones, with the drive shafts acting as upper links, additional location being provided by radius arms, and springing by twin coil/damper units.

There is remarkably little roll during cornering, and the balance is excellent, though it understeers quite strongly in relatively tight bends. At higher speeds it is more neutral, and will eventually oversteer, though it is doubtful if many owners will reach that point. Roadholding on the big Dunlop tyres with which all XJs have up to now been fitted is very good, if not outstanding, and the car remains safe and easy to handle even on wet roads. This is not, however, to suggest that it is devoid of faults.

Regular readers will know that one of our few criticisms of the XJ range has consistently concerned what in our view has been the excessive assistance of the rack-and-pinion power steering. In the new car, Jaguar have gone some way towards assuaging our feelings: there is more weight and slightly more feel, and the "over-centre" sensation when lock was applied (discernible by a loss of weight when the wheel was turned from the straight-ahead position) has been eliminated.

However, we are still not greatly impressed by the feel imparted through the steering wheel, and on "our" car straightline stability on an uneven country road at speed was not as good as it has been in previous XJ6s.

MOTOR ROAD TEST No. 42/79

★★★★★ excellent | ★★★★ good | ★★★ average | ★★ poor | ★ bad

Make: Daimler
Model: Sovereign 4.2
Maker: Jaguar Rover Triumph Ltd, Browns Lane, Allesley, Coventry, West Midlands, CV5 9DR
Price: £11,652.00 plus £971.00 car tax plus £1.893.45 VAT equals £14,516.45. Extras fitted to test car: air conditioning (£1051.59), electrically-operated steel sunroof (£491.15), cruise control (£272.85), stereo radio/cassette player (£199.25), headlamp wipe/wash (£174.63), electrically-operated height adjustment of driver's seat (£159.72), electrical adjustment of door mirrors (£98.22), front fog lights (£77.45). Price as tested: £17,041.31

BRAKES

★★★★

Our brake fade test (devised in collaboration with Girling) is a severe one, but we expected the Daimler's massive Girling discs, which are ventilated at the front, to survive it without problems. However, on the 19th stop from 0.6 g, the system faded totally, the pedal going right to the floor because the fluid had boiled.

Jaguar were equally surprised, and extremely concerned, so in our presence the brakes were bled and we carried out the test again. This time the car passed with flying colours, required pedal pressure hardly showing any variation apart from a slight drop once the pads had warmed up. We have no reason to dispute Jaguar's conclusion about its former failure, that contaminated fluid had been supplied.

Even before the fluid had been changed, we did not experience even a suggestion of fade on the road, despite some fairly hard driving over a demanding route. Slowing this big car is at one time smooth and progressive, and reassuringly powerful. With only 68 lb pressure, it is possible to achieve more than 1.0 g deceleration from 30 mph.

The handbrake of our test car (still of the umbrella type, unlike the XJS) was uncharacteristically weak, only managing a 0.34 g emergency stop, and being unable to hold the car downhill on the 1-in-4 test slope.

ACCOMMODATION

★★★★

One of the greatest advantages of the Series III's modified body line is extra headroom for rear-seat passengers, and the greater glass area gives an even more airy feel to the car. There are few more civilised forms of transportation for four full-sized adults, and for those who do not like leather seats (in this case, high quality Connolly hide), cloth trim is optional at no extra cost. Even with the front seats set on the rearward extremity of their runners, there is plenty of knee room in the back of the car.

Internal stowage is provided by a locker on the passenger's side of the facia, a lidded cubby in the centre console (suitable for carrying cassettes), a small pocket in each of the front doors, a tray on the facia, pockets in the backs of the front seats, and the rear parcel shelf. The boot, which holds 10.9 cu ft of our standard Revelation suitcases, is long, wide and easy to load, but shallow.

RIDE COMFORT

★★★★

One of the comments we have made about the XJ's ride quality in the past is that what you hear emphasises what you feel, and one great achievement of Jaguar's engineers has been the elimination of many of the noises induced by road surfaces which are noticeable in even the most expensive versions of cars built by Rolls-Royce and Mercedes-Benz.

The Series III seems to be a little firmer than its predecessors, but it shares with them a virtual absence of tyre thump over bumps and hollows. The ride is uncannily smooth over badly broken surfaces, and excellent at high speed on motorways and fast country roads, though there is a trace of float after humps or severe undulations. Where it does not seem to perform quite so well as before is at low speed over transverse ridges, and there is noticeable jiggling on bumpy suburban roads.

AT THE WHEEL

Drivers of all shapes and sizes should fit comfortably into the Daimler. The well shaped seat is adjustable fore and aft, for backrest angle, for lumbar support and (at extra cost) for height (the latter controlled by an electric switch), and the steering wheel can be adjusted for reach; unfortunately, when the wheel is pulled towards the driver, the stalks stay where they are. The throttle and brake pedals are well placed, leaving plenty of room to rest the left foot, and all controls are within easy reach.

Sited to the right of the wheel, the wiper stalk has three settings, two for speed and one for intermittent, and a button at the end which operates the windscreen wash (also the optional headlamp wipe/wash when the lights are on). The left-hand stalk operates the indicators and dip switch/flash, with a button on the end for setting the cruise control (another option); to resume a chosen cruising speed after the driver has disengaged the cruise control by braking, there's a rocker switch behind the gear selector lever. Switches for the electrically-operated windows and sunroof are laid out in a logical pattern on the forward-facing section of the centre console, while most other minor controls (of the push-push button variety used by Jaguar/Daimler for some time) are set on the facia, to the left of the main lights switch. Ashtrays are provided on either side of the gear lever.

VISIBILITY

The increased rake of the windscreen has not affected the excellent forward visibility, and all round, the new car is slightly easier to see out of than the model it replaces, thanks to a greater area of glass and slimmer rear pillars. Some drivers mentioned that it was rather difficult to reverse accurately, owing to the drop of the rear wings.

The interior mirror has been enlarged and this coupled with the big door mirrors (adjustable on our test car electrically from "joystick" switches on the driver's door) give a comprehensive view of following traffic.

Quartz halogen headlamps, previously an extra-cost option, are now standard equipment, and we found them powerful on both dipped and main beams.

INSTRUMENTS

Eschewing modern trends, Jaguar continue to fit their cars with vertically-faced instruments mounted in a highly polished walnut facia. The instrumentation is comprehensive, the dials clearly marked and very accurate, with the exception of the fuel gauge. The large rev counter and speedometer (separated by a bank of warning lights) are flanked on the left by an oil pressure gauge and a voltmeter, and on the right by the fuel gauge (which shows the level in only one of the twin tanks; a switch on the facia controls which tank feeds the engine and is monitored by the gauge) and water temperature gauge. A small quartz clock is set beneath the tray in the middle of the facia.

We imagine that most owners will find the instrument design in the Series III attractive, and not worry that by comparison with some rivals (notably BMW) it is old fashioned. But it is indisputable that on sunny days reflections can render some or all of the dials unreadable. Other manufacturers have adopted instrument arrangements with a sloping glass that eliminates all reflections. We see no reason why such a concept could not be incorporated in a facia which would still have a traditional English flavour.

AIR CONDITIONING

Our test car was fitted with the optional air conditioning system (£1052 extra). It has two major controls, at either side of the centre console; the right hand dial has four settings, Low, Automatic, High and Defrost. The left hand dial controls temperature, with a range of 65° to 85°F. In the Defrost position, the automatic facility is overriden, and airflow at maximum temperature is provided, 90 per cent directed at the windscreen and the remainder to the footwells and rear passenger compartment. The screen is cleared almost immediately, even when thoroughly misted up.

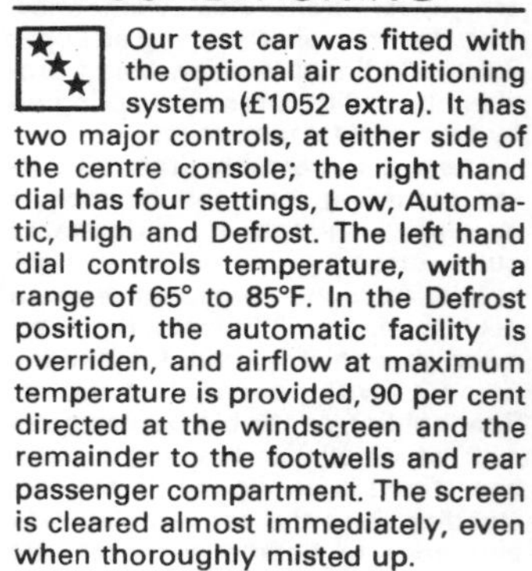

On Auto, the control system senses internal and external temperatures in relation to the setting on the left hand dial, and then adjusts flaps and heating or cooling modes accordingly. The High and Low settings override the fan speed setting.

On a hot day, the throughput of cooled air is excellent, and the direction of cooled air to the face is easily adjusted by the centre vents on the facia. This flow can be supplemented by that from the facia end vents. However, although the conditioner has been improved (it reacts more quickly to adjustments), it suffers from a major fault which afflicts most refrigeration systems, with the notable exception of that in Rolls-Royces: on a cold or cool day, it is not possible to obtain ventilation and footwell heating simultaneously, so that although the heat output is high it can take a long time before a comfortable atmosphere is achieved.

NOISE

If we had to single out the quality which contributes most to the appeal of the XJ range of saloons, it would be refinement: there is no other car we have tested, at any price, which is quieter than the V12 cars at a given speed, and the straight-six models are almost as good.

The current 4.2 Sovereign is no exception to this excellence. Uncannily quiet at low speed and when cruising, with just a trace of rumble from the big tyres as they move over surface irregularities, its noise suppression is better than ever in some areas. Induction roar is definitely reduced, though there is still a slight throbbing from the engine as 5,000 rpm is approached. Wind noise is virtually undetectable below about

Right: the air conditioning controls are set on either side of the radio/cassette player, which is in this case the optional Philips 880 stereo unit. Above: the facia (seen through the new steering wheel) is unchanged; it has strong showroom appeal, but reflects badly

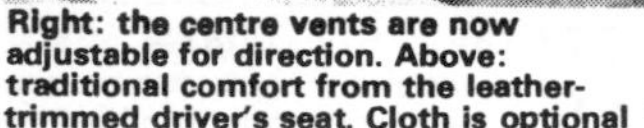

Right: the centre vents are now adjustable for direction. Above: traditional comfort from the leather-trimmed driver's seat. Cloth is optional

Above: there is more room than ever before in the rear of the XJ

Left: this velour-covered oddments tray on the facia is useful for cigarette packs or cassettes

MOTOR ROAD TEST NO 42/79 ● DAIMLER SOVEREIGN

Above: easily loaded boot held 10.9 cu ft of test luggage. Below: the lefthand stalk sets the cruise control

80 mph, rising steadily thereafter but never getting anywhere near irritation level. The readings with our noise level meter confirmed our subjective impressions — the maximum reading we obtained (72 dBA at 5,000 rpm with the transmission held in bottom gear) was lower than many cars cruising at 50 mph, an outstanding achievement.

Part of the reason for this improvement (and it is heartening to discover that the company is not sitting on its laurels) is improved insulation built into the new bodyshell, mainly in the areas of the bulkhead, doors and transmission tunnel.

FINISH

The bodyshells of the Series III XJs are built at the Pressed Steel Fisher plant at Castle Bromwich, where what is described as Britain's best paint shop has recently been completed. There is a new range of nine colours from which to choose. The finish of our test car was immaculate, and we particularly liked the alterations which have been carried out, in particular the new bumpers and flush-fitting door handles, which not only improve the appearance of the car but should also keep it looking good for a longer time. The smart new wheel trims are retained by the wheel nuts, thus making wheel changes simpler.

The interior is as beautifully finished as ever, and most owners will like its traditional appearance. Higher quality carpets have been used and the rooflining has been modified. Otherwise it's very much as it's always been, and as always, we very much approve.

EQUIPMENT

The Sovereign's price will seem high to many people, but in relation to its fittings and to the offerings of the opposition, it follows a Jaguar tradition of outstanding value for money.

For your £14,516, you get electric

continued over

PERFORMANCE

CONDITIONS

Weather	Sunny, wind 0-10 mph
Temperature	72°F
Barometer	29.5 in Hg
Surface	Dry tarmacadam

MAXIMUM SPEEDS

	mph	kph
Mean of two autobahn runs	128.0	206.0
Terminal Speeds:		
at ¼ mile	81	130
at kilometre	102	164
Speed in gears (at 5000 rpm):		
1st	52	85
2nd	85	137

ACCELERATION FROM REST

mph	sec	kph	sec
0-30	4.2	0-40	3.4
0-40	5.9	0-60	5.4
0-50	7.7	0-80	7.6
0-60	10.5	0-100	11.1
0-70	13.5	0-120	14.9
0-80	16.9	0-140	20.1
0-90	21.6	0-160	27.5
0-100	28.4		
0-110	38.1		
Stand'g ¼	17.6	Stand'g km	31.8

ACCELERATION IN KICKDOWN

mph	sec	kph	sec
20-40	3.5	40-60	2.1
30-50	4.1	60-80	2.8
40-60	5.5	80-100	3.7
50-70	5.7	100-120	3.6
60-80	6.0	120-140	5.1
70-90	8.2	140-160	8.3
80-100	11.8		
90-110	16.5		

FUEL CONSUMPTION

Touring*	19.3 mpg 14.6 litres/100 km
Overall	15.7 mpg 18.0 litres/100 km
Govt tests	14.0 mpg (urban) 27.7 mpg (56 mph) 21.9 mpg (75 mph)

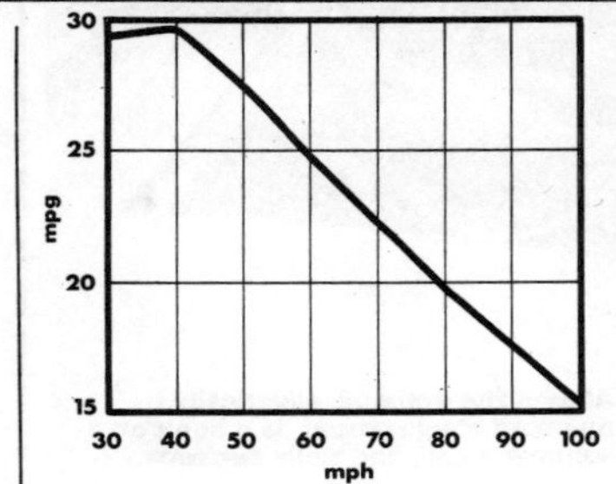

Fuel grade	97 octane 4 star rating
Tank capacity	20 galls 45.5 litres
Max range	386 miles 621 km
Test distance	1293 miles 2080 km

*Consumption midway between 30 mph and maximum less 5 per cent for acceleration.

BRAKES

Pedal pressure, stopping distance, and average deceleration from 30 mph (48 kph).

lb	kg	ft	m	g
25	11.4	77.0	23.5	0.39
50	22.7	56.0	17.1	0.54
68	30.9	30.0	9.2	1.04
Handbrake		88	26.8	0.34
Maximum from 70 mph (113 kph)				
69	31.3	173	53	0.96

FADE

Twenty 0.6 g stops at 45 sec intervals from speed midway between 40 mph (64 kph) and maximum (83 mph, 134 kph) at gross vehicle weight.

	lb	kg
Pedal force at start	44	20.0
Pedal force at 10th stop	38	17.3
Pedal force at 20th stop	38	17.3

STEERING

Weighting at wheel rim when parking and when cornering on 108 ft diameter circle.

	lb ft
Parking	4.0
Cornering at 0.1 g	3.0
0.3 g	3.0
0.6 g	2.5

Turning circle between kerbs	ft	m
Left	39.1	11.9
Right	39.7	12.1
Lock to lock	3.2 turns	
50 ft diam circle	1.3 turns	

NOISE	dBA	Motor rating*
30 mph	56	6
50 mph	63	11
70 mph	69	16
Max revs in 1st	72	18

*A rating where 1 = 30 dBA and 100 = 96 dBA, and where double the number means double the loudness.

SPEEDOMETER (mph)

Speedo							
30	40	50	60	70	80	90	100
True mph							
30	40	50	60	70	80	90	100

Distance recorder: 100 per cent accurate

WEIGHT

	cwt	kg
Unladen weight*	35.4	16.1
Weight as tested	39.1	17.8

*with fuel for approx 50 miles

Performance tests carried out by Motor's staff at the Motor Industry Research Association proving ground, Lindley.

GENERAL SPECIFICATION

ENGINE

Cylinders	Six in line
Capacity	4235 cc (258 cu in)
Bore/stroke	92.07/106 mm (3.625/4.173 in)
Cooling	Water
Block	Cast iron
Head	Aluminium
Valves	Dohc
Cam drive	Chain
Compression	8.7:1
Induction	Lucas/Bosch electronic injection via water-heated aluminium inlet manifold
Bearings	7 main
Max power	205 bhp (DIN) at 5000 rpm
Max torque	232 lb ft (DIN) at 4500 rpm

TRANSMISSION

Type	Borg Warner 65 3-speed automatic with torque converter
Internal ratios and mph/1000 rpm	
Top	1.00:1/24.7
2nd	1.46:1/16.9
1st	2.40:1/10.3
Rev	2.09:1
Final drive	3.07:1

BODY/CHASSIS

Construction	Integral, all steel
Protection	Multi-stage phosphate spray and dip, anti-corrosive electoprimer, "adhesion promotor", thermoplastic acrylic paint, oil sanding, oven treatment; wax injected box sections; full underbody sealant

SUSPENSION

Front	Independent, by semi-trailing wishbones and coil springs, with anti-dive geometry; anti-roll bar
Rear	Independent, by lower wishbones, fixed-length driveshafts acting as upper links, radius arms, twin coil springs

STEERING

Type	Rack and pinion
Assistance	Yes

BRAKES

Front	Ventilated disc, 11.18 in diameter
Rear	Disc, 10.38 in diameter
Park	On rear wheels
Servo	Yes
Circuit	Dual, split front/rear
Rear valve	Yes
Adjustment	Automatic

WHEELS/TYRES

Type	Steel, 6J×15
Tyres	Dunlop SP Sport ER 70 VR 15
Pressures	27/26 psi F/R (up to 100 mph) 27/30 psi F/R (up to 100 mph, full load) 33/32 psi F/R (high speed)

ELECTRICAL

Battery	12V, 66 Ah
Earth	Negative
Generator	60Ah alternator
Fuses	22
Headlights	
type	Quartz halogen
dip	55 W total
main	115 W total

GUARANTEE

Duration ..12 months, unlimited mileage

MAINTENANCE

Free service: ....at 1000 miles
First main service: ....at 3000 miles
Schedule: ....every 6000 miles

DO-IT-YOURSELF

Sump: ....12 pints, 20W50
Gearbox .16 pints, automatic fluid type F
Rear axle ....2.75 pints, SAE 90
Coolant: ....32 pints
Chassis lubrication: ....none
Spark plug gap: ....0.035 in
Spark plug type ....Champion N11Y
Tappets: (hot) ....0.012-0.014 Inlet
......0.012-0.014 Exhaust

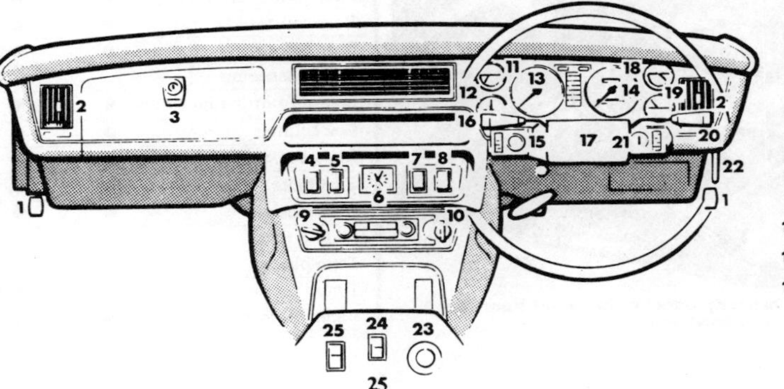

1 footwell vents
2 facia vents
3 glove locker
4 petrol tank changeover switch
5 heated rear window switch
6 clock
7 map light switch
8 interior light switch
9 heater volume control
10 temp and vent control
11 oil pressure gauge
12 batt. cond. indicator
13 tachometer
14 speedometer
15 lights switch
16 indicators/headlamp flash and dip
17 horn push
18 fuel gauge
19 water temp gauge
20 wash/wipe/delay arm
21 ignition/steering lock
22 bonnet release
23 cigar lighter
24 cruise control
25 central locking button

window lifters, central locking, leather upholstery, and a radio/cassette player with an electrically operated aerial. This is in addition to a complex mechanical specification which includes fuel injection, and Lucas Opus transistorised ignition.

Our test car was fitted with several optional extras: the electrically-operated steel sunroof (£491.15), electric adjustment of the door mirrors (£98.22), and cruise control (£272.85), front fog lights (£77.45), electrical adjustment of the driver's seat height (£159.72), a stereo radio (£199.25), headlamp wipe/wash (£174.63), and air conditioning, which has already been mentioned.

IN SERVICE

AFTER the first (free) service at 1,000 miles, Jaguar recommend a maintenance interval of 6,000 miles, at which the main job is changing the oil, apart from an initial check-up after the first 3,000 miles have been covered.

None of the cars in this league is cheap to run, but many replacement parts are cheaper on the Jaguar/Daimler range than the equivalent parts for its main rivals, the BMW and Mercedes-Benz.

The front-hinged bonnet, released by a lever near the driver's door, is self-supporting. The underbonnet view, though tidy, is rather daunting to the man who likes to carry out maintenance work himself, but the battery, fluid levels, and sparking plugs are readily accessible. There are two fuse boxes, one on each side of the facia.

Apart from the jack and wheelbrace, a useful set of tools is supplied in a neat little attache case clipped to the wall of the boot. This contains five open-ended spanners, a tyre pressure gauge, a pair of pliers, a manual winder for the sunroof in the event of electrical failure, four replacement bulbs, and three fuses.

IN CONCLUSION

THERE is an air of renewed enthusiasm at Jaguar Rover Triumph since it has once again become more of a separate division within BL Cars, and such enthusiasm is reflected in an improved product range. Though still competitive, the Series II was overdue for further development, and in most respects we approve wholeheartedly of the changes that have been made.

Above: the optional, electrically operated steel sunroof, is a boon on a summer's day, and seals perfectly.

Below: rear seat passengers must fight for directional control of this large eyeball vent

Not only is this 4.2-litre version quicker and more refined than before, it is also significantly more economical, which is exactly what is required for today's more expensive motoring.

The revised bodywork and exterior trim, typical of Jaguar's conservative and methodical approach, is more elegant and more practical than before, and a major advance has been the increase in headroom for back-seat passengers.

It is a pity that some of the long-standing failings of the car — though many of them seem like nit-picking when set against the XJ's many virtues — could not have been eradicated. For example, we would still like to have more weight in the steering, and the instrument display (handsome though it is in the showroom) can be all but unreadable on a sunny day.

Keeping on the Daimler name, though only in badge and grille engineering, is sensible, as it retains a lot of prestige. This is, in many ways the best saloon car in the world, and most certainly the best value for money in its price range in this country.

Above: now injected, the 4.2-litre engine is neatly installed, but apart from very minor jobs is beyond the abilities of the home handyman

Comparisons

PERFORMANCE	Daimler	BMW§	Cadillac	Ford§	Mercedes	Opel†
Max speed, mph	128	128**	103.2	117.4	122.3	125**
3rd	—	98	—	91	—	—
2nd	85	62	94	66	85	90
1st	52	35	56	40	45	56
0-60 mph, secs	10.5	8.3	12.6	9.0	10.4	9.4
30-50 mph in 4th††, secs	4.1	8.5	4.3	9.3	5.0	3.4
50-70 mph in top, secs	5.7	8.3	8.4	10.1	5.9	5.2
Weight, cwt	35.4	31.6	39.3	26.1	33.7	27.3
Turning circle, ft*	39.3	35.5	41.7	32.2	32.8	29.8
50ft circle, turns	1.3	1.2	1.1	0.7	1.0	1.1
Boot capacity, cu ft	10.9	12.6	8.0	13.2	15.0	10.7

§figures from manual gearbox versions †figures from Opel Monza
**estimated ††kickdown for automatics
*mean of left and right

COSTS AND SERVICE	Daimler	BMW	Cadillac	Ford	Mercedes	Opel
Price, inc VAT & tax, £	14516	15011	18232	8883	16840	11260
Insurance group	8	8	7	6	7	7
Overall mpg	15.7	17.6	12.4	20.4	13.5	17.4
Touring mpg	19.3	—	—	—	—	—
Fuel grade (stars)	4	4	2	4	4	4
Tank capacity, gals	20.0	18.7	17.5	14.6	24.2	15.4
Service interval, miles	6000	5000	6000	6000	6000	5000
No of dealers	319	150	1	1200	98	230
Set brake pads (front) £*	23.33	20.50	33.86	17.87	16.59	18.96
Complete clutch £*	—	114.06	—	58.94	—	—
Complete exhaust £*	255.86	212.82	419.75	80.62	196.62	218.21
Front wing panel £*	110.40	98.34	332.72	50.14	81.60	52.51
Oil filter, £*	5.75	2.60	7.97	4.31	2.86	5.52
Starter Motor, £*	78.87	100.79	191.99	67.47†	91.43†	143.48
Windscreen, £*	78.20**	84.68**	270.96**	48.10**	132.91**	140.30**

*inc VAT but not labour charges †exchange **laminated

STANDARD EQUIPMENT	Daimler	BMW	Cadillac	Ford	Mercedes	Opel
Adjustable steering	●	●	●		●	●
Air Conditioning			●			
Alloy Wheels		●		●		●
Central door locking	●	●	●	●	●	●
Cigar lighter	●	●	●	●	●	●
Clock	●	●	●	●	●	●
Cloth trim	I	●	●	●	I	●
Dipping mirror	●	●	●	●	●	●
Driver seat height adjust		●	●		●	●
Driver seat tilt adjust	●	●	●	●	●	●
Electric window lifters	●		●	●	●	●
Fresh air vents	●	●	●	●	●	●
Headlamp washers				●		●
Head restraints	●	●	●	●	●	●
Heated rear window	●	●	●	●	●	●
Intermit/flick wipe	●	●	●	●	●	●
Laminated screen	●	●	●	●	●	●
Locker	●	●	●	●	●	●
Passenger door mirror	●		●	●		
Petrol filler lock	●	●				●
Power steering	●	●	●	●	●	●
Radio	●		●	●		●
Rear central armrest	●	●	●	●	●	●
Rear courtesy light	●	●	●		●	
Rear fog light	●	●			●	●
Rear wash/wipe						
Remote mirror adjustment	●	●	●	●	●	●
Rev counter	●	●		●	●	●
Reverse lights	●	●	●	●	●	●
Seat belts — Rear	●	●				●
Seat recline	●	●		●	●	●
Sliding roof				●		
Tape player			●	●		●
Tinted glass	●	●	●	●	●	●
Vanity mirror	●	●	●	●	●	●

The Rivals

The Jaguar/Daimler range remains good value compared with its opposition, since most rivals are slower, less well equipped, or more expensive — and in some cases a combination of all three

DAIMLER SOVEREIGN SIII 4.2 £14,516

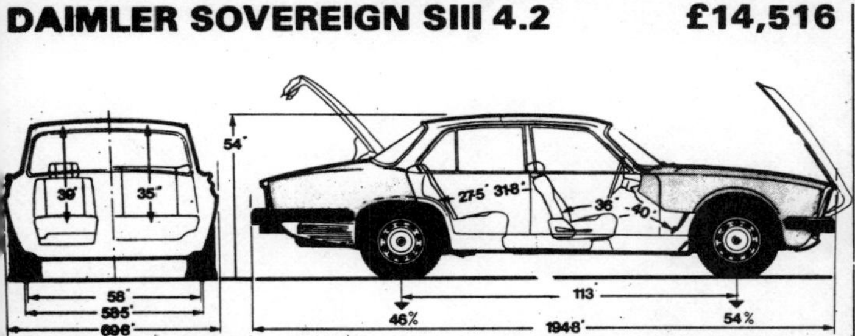

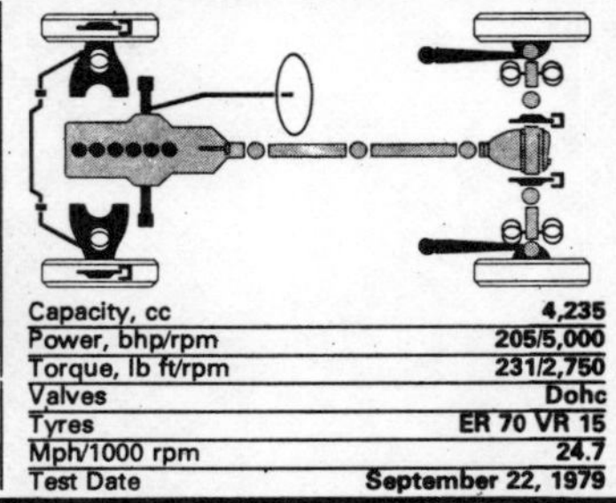

Capacity, cc	4,235
Power, bhp/rpm	205/5,000
Torque, lb ft/rpm	231/2,750
Valves	Dohc
Tyres	ER 70 VR 15
Mph/1000 rpm	24.7
Test Date	September 22, 1979

The long-awaited Series III XJ range incorporates many important detail improvements. The 4.2-litre versions are now injected, faster and more economical than ever before, and also even more refined. The appearance is neater, and there is more headroom. Ride remains very good but the steering is still too light, and firmer damping would be advantageous. Despite this and other remaining faults, it remains an eminently civilised machine.

BMW 733i AUTO £15,011

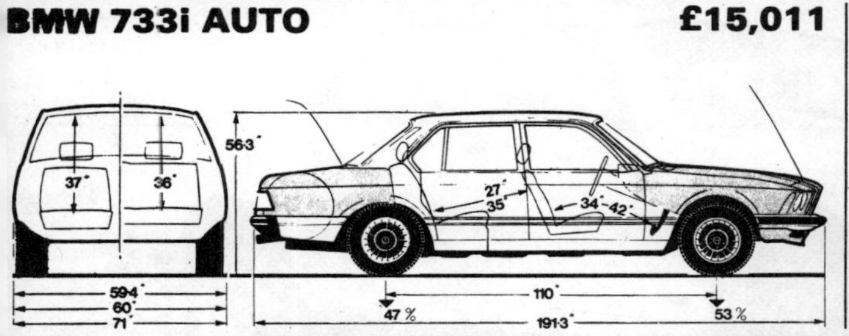

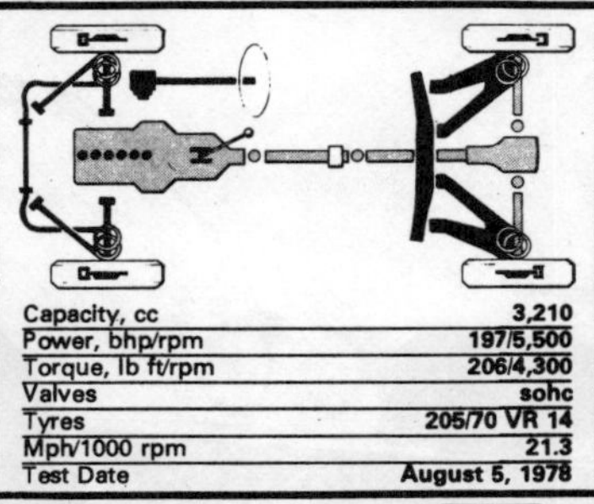

Capacity, cc	3,210
Power, bhp/rpm	197/5,500
Torque, lb ft/rpm	206/4,300
Valves	sohc
Tyres	205/70 VR 14
Mph/1000 rpm	21.3
Test Date	August 5, 1978

Top model in BMW's four door 7-series range. Features 3.3-litre fuel injected engine which gives this big saloon an excellent performance but at the expense of fuel consumption. More expensive than most cars of this capacity but it does offer (in addition to performance) good road manners, except when laden, a superb finish, many extras and plenty of room. Not as much fun as old 3.3 cars, but more refined.

CADILLAC SEVILLE £18,232

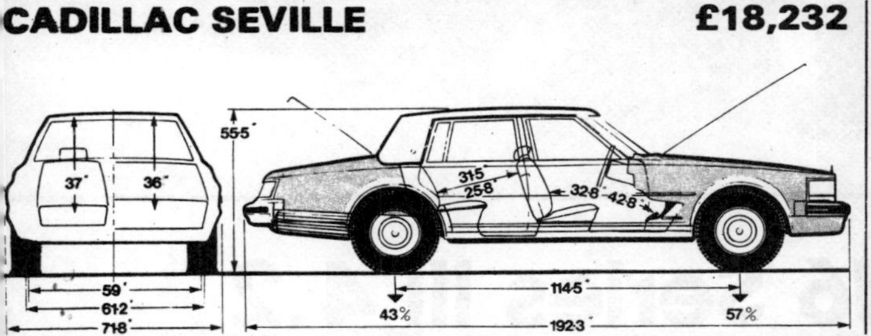

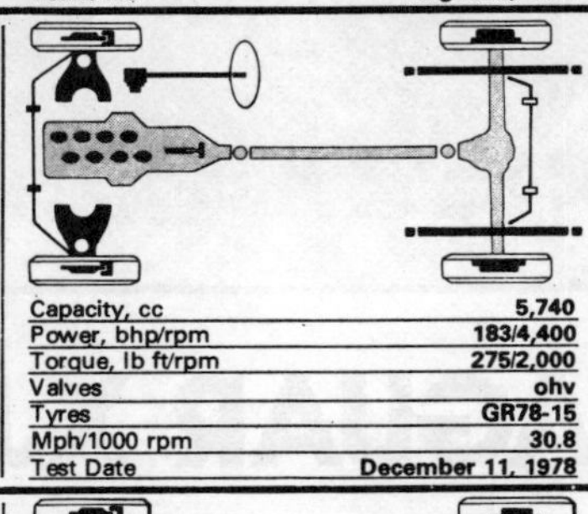

Capacity, cc	5,740
Power, bhp/rpm	183/4,400
Torque, lb ft/rpm	275/2,000
Valves	ohv
Tyres	GR78-15
Mph/1000 rpm	30.8
Test Date	December 11, 1978

Small by American standards, large by European ones, the Cadillac is very much a horse for an American course, comfort, quietness and smoothness being second to none on smooth straight roads. It does not like corners, though, or indifferent surfaces. With its self-dipping lights, and chime to remind you to fasten seat belts, the Cadillac provides a novel experience for British drivers but at a high price.

FORD GRANADA 2.8 GHIA AUTO £8,883

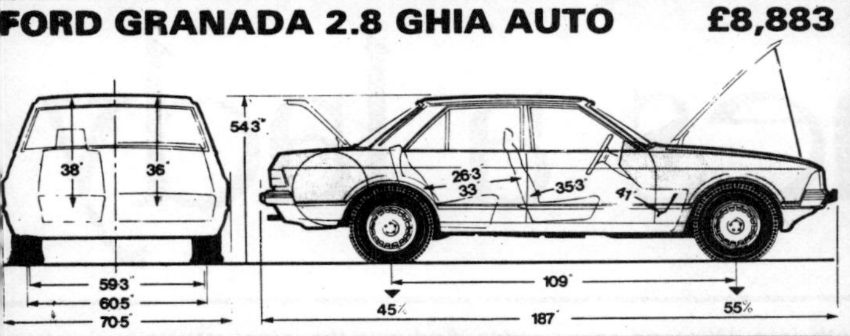

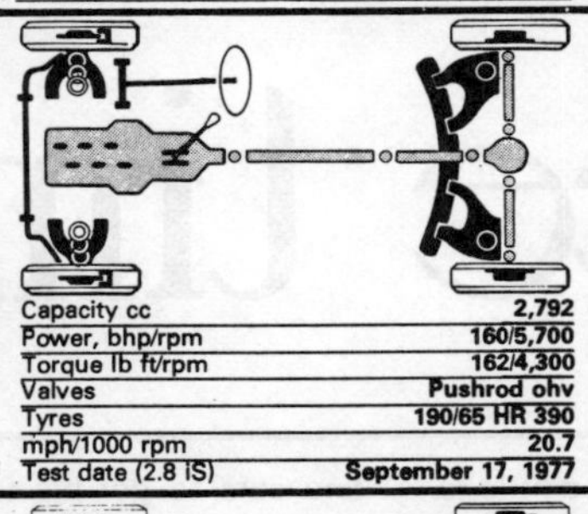

Capacity cc	2,792
Power, bhp/rpm	160/5,700
Torque lb ft/rpm	162/4,300
Valves	Pushrod ohv
Tyres	190/65 HR 390
mph/1000 rpm	20.7
Test date (2.8 IS)	September 17, 1977

Most expensive, luxury version of Ford's Granada saloon, though tested by us in similiar iS manual form, with good performance though at the expense of low-speed torque. Good ride, handling and roadholding, and generally very refined. Comfortable, plush interior, very roomy, and very well equipped — even a sunroof is standard. Brakes a bit suspect under hard use, and steering lacks feel.

MERCEDES-BENZ 350SE £16,840

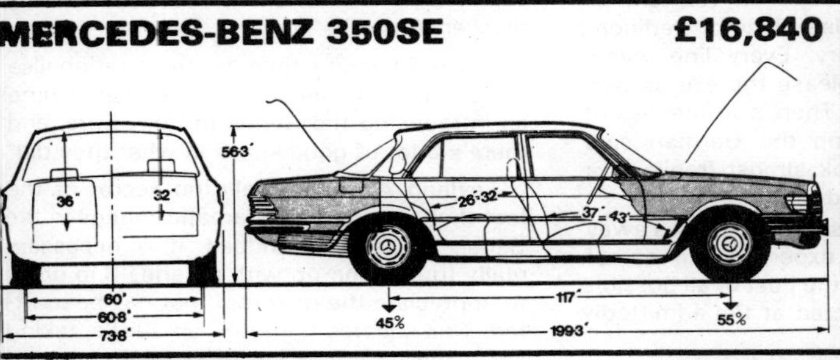

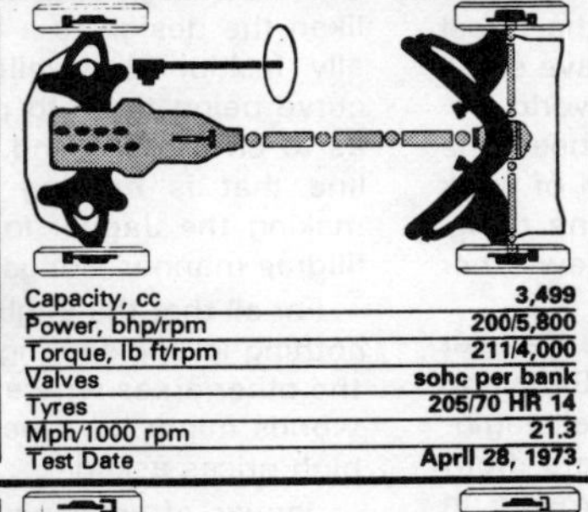

Capacity, cc	3,499
Power, bhp/rpm	200/5,800
Torque, lb ft/rpm	211/4,000
Valves	sohc per bank
Tyres	205/70 HR 14
Mph/1000 rpm	21.3
Test Date	April 28, 1973

3.5-litre V8-engined version of short-chassis S-Class Mercedes. A magnificent car, with superb roadholding and handling. Other plus points include high comfort levels, a smooth automatic transmission, good high-speed cruising capabilities, many built-in safety features and it is beautifully made and finished. However it is not especially quiet, economy is poor, the ride is firm at low speeds, and it is pricey.

OPEL SENATOR £11,260

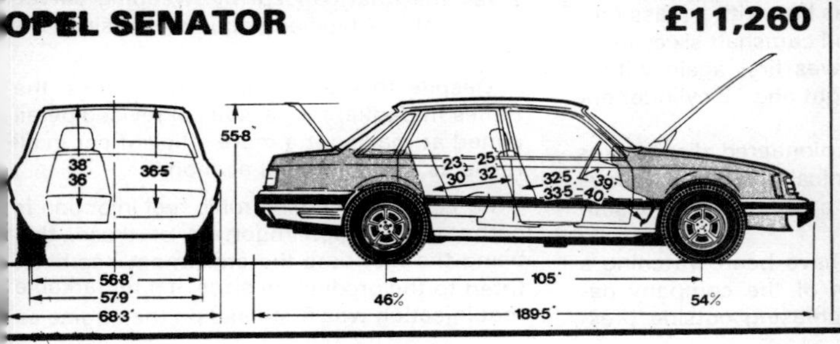

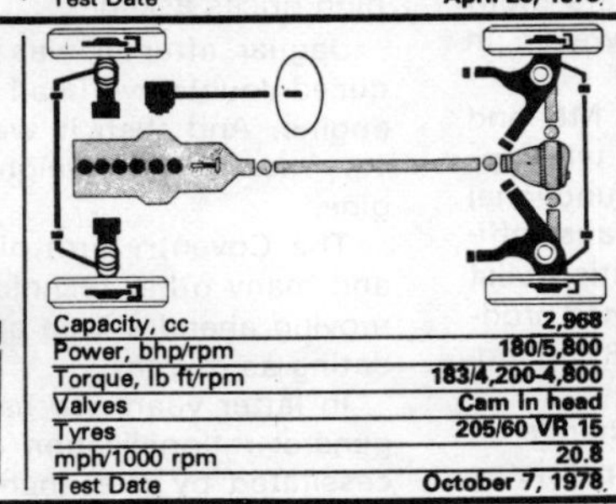

Capacity, cc	2,968
Power, bhp/rpm	180/5,800
Torque, lb ft/rpm	183/4,200-4,800
Valves	Cam in head
Tyres	205/60 VR 15
mph/1000 rpm	20.8
Test Date	October 7, 1978

Opel's prestige saloon combines refinement and comfort with brisk performance. Smooth-changing automatic gearbox (manual available to special order), excellent ride and handling, good brakes; luxurious, lavishly equipped, though not as spacious as some rivals. Good heating, excellent ventilation, huge boot. Some aspects of the interior finish are not to our taste, and the engine is a bit thirsty. Plenty of appeal for the sporty driver.

FULL TEST **JAGUAR XJ6 Series III 4.2**

Three times a lady

WHILE IT MAY be stretching things a little to refer to Jaguars continuously as the "Best cars in the world," they certainly have some features which are the best in the world.

Silence of operation has always been one of these strong points, a great deal of work having gone into achieving that one thing. With it goes a smoothness which few other makes have been able to duplicate.

In our book, Jaguar (including the Daimler range) is on a par with Mercedes Benz and BMW in producing sophisticated well engineered luxury automobiles. It differs from both its German counterparts, however, in total philosophy.

Teutonic efficiency makes both MB and BMW sheer delight to the person who appreciates such modern works of functional art. They look the way they are because efficiency dictated it that way, the artistic stylist only getting a minor role in the finished product. And simply because they ARE good, they LOOK good.

In the case of Jaguar however, there's far more soul for the person who appreciates fine cars to get his teeth into. We tend to liken the design to a fine piece of traditionally fashioned jewellery. Every line, every curve being there to please the eye as well as to cheat the wind. There's a fineness of line that is missing on the German cars making the Jaguar look almost fragile in a filigree manner alongside.

For all that, the English marque gives away nothing in engineering expertise or in any of the other areas where the best of all possible worlds might be expected at the admittedly high prices asked.

Jaguar, after all, was first with a mass produced double overhead camshaft six cylinder engine. And then it was first again with a very modern, lightweight ohc 12 cylinder engine.

The Coventry firm pioneered disc brakes and many other technical advances, always moving ahead a little at a time and consolidating as it went.

In latter years we have been watching a general rationalisation of the company necessitated by ever increasing outside pressures. And now the latest rationalisation has reached Australia.

It was as good a time as any to rationalise the Jaguar Daimler range at the same time as introducing the Series III body here, and there's a lot of good sense in what they did.

Leyland Australia make no secret of the fact that there is little dramatic change in the Series III Jaguars. Indeed, it is necessary really to line one up with a Series II in order to appreciate the new roof that has been fitted. This squares the looks up a little, taking away the total concept of sweeping curved lines that has typified Jaguars ever since the Mk 5.

Despite this lack of innovative looks, the Series III Jaguar range is full of revised detail aimed at making it a more efficient car in all respects, including fuel economy.

The Lucas Bosch electronic fuel injection is perhaps the most important of these, this being the first time the equipment has been fitted to the production sixes. It's remarkable that injection was first used on the engine as

The Jaguar XJ series has been called the best saloon car in the world. Now the third version has been released, Jaguar has improved on the thoroughbred breed even further. Paul Harrington says the Series III is closer to the title of the "world's best" than anything previously produced by the people from Coventry.

long ago as 1957 when the D Type sports racing models achieved their incredible 1, 2, 3, 4, and 6th places at Le Mans. It has taken a long time for it to be adopted on other than the 5.3 litre V12 that's for sure!

The headlights are now of halogen type rather than the sealed beam tungsten fitted before.

Revised seats have wider section pleats and 38mm added to the seat squab height, as well as revised headrests. Lumbar support adjustment is now standard on the front seats while, on the V12 engined Daimlers it is possible to obtain electric height adjustment for the driver and front passenger seats. Rear seat room has been improved to provide an extra 38mm in height for the taller passenger.

Then there's the new thicker tufted cut-pile carpeting, together with revised noise deadening underneath.

At long last there's an intermittent control for the windshield wipers in addition to all the other speeds available and the "flick" single wipe mode.

Larger rear view mirrors improve safety, and there's an interior light delay to allow passengers to settle in before the light goes out.

Even the car's locking system has been updated. It's now a three key system. A master key will lock all four doors and the boot with one movement. Thereafter a separate "service" key is necessary to open the doors but this will leave the boot locked. Only the master key will unlock this. When the car is taken in for service therefore only the service key needs to be handed over, the security of the locked boot allowing valuables to be left inside without worry. A third key is for the ignition alone.

Other new bits include door handles, steering wheel and trim parts. Sadly neglected to date, there's also a radio/stereo cassette player fitted as STANDARD across the range. This is the superb KEX-20. We feel sure that Jaguar could argue justifiably that they waited until there was a sound system of sufficiently high standard to compliment their engineering before making it standard. In which case they're not going to get any argument from us! It features its own separate amplifier which is hidden away inside the console mounting. There's also Dolby noise reduction and a chrome tape switch.

Although it first saw the light of day as long ago as 1948, the Jaguar six cylinder twin overhead camshaft engine has undergone continual development. In fact, the current engine has very little in common with the original, having different spacing for the bores among other things. The addition of fuel injection has increased power output by around 30 kW, making it more of a hot performer than before. More importantly these days though, fuel economy has been improved, especially on the three speed auto which, we understand from our European correspondent, is a little more economical even than the manual. We would suggest that this stems from the likelihood that the five speeder would be driven in a more sporting manner however!

Top: The superb six is now restricted to the Jaguar saloons, the 12 being fitted to the even more exclusive Daimler.
Above: Leather bound seats are comfortable in the Jaguar fashion, and overall quality of finish, as you would expect, is very high.
Below: Big back bench seat is equally comfortable and there's a surprising amount of legroom back there too.

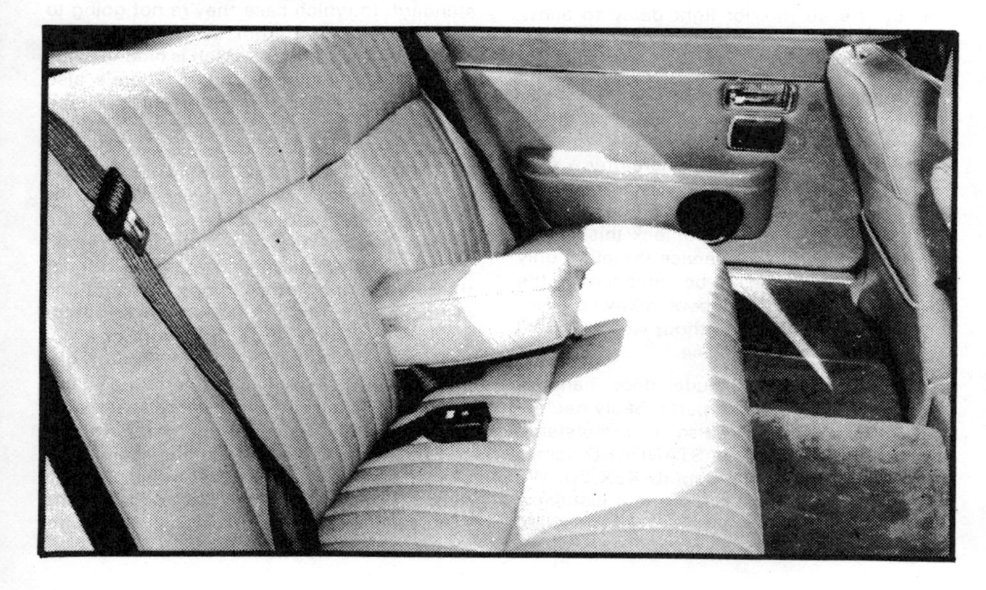

It's a traditionally long stroke engine which runs on an 8.1:1 compression ratio, producing 152.9 kW DIN at 5000 rpm. There is also 314 Nm of torque at 1500 rpm.

The cast iron block is fitted with interference fit dry iron liners. The crank runs in seven main bearings and is a molybdenum steel forging with three plane configuration.

Aluminium is used for the cylinder head with its twin cams, this being the first mass produced twin cam six cylinder at its introduction forty years ago.

Price of the XJ 4.2 has risen with the introduction of the Series III going from $32,500 to $38,500. Remember however, that the new car gets the injection and the Pioneer stereo set-up as well as all the other refinements described above.

The Daimler Sovereign, again 4.2 litre six cylinder engined, rises from $33,480 to $39,500, while the Double six with its V12 engine of 5.3 litres goes from $34,540 to $44,500 with even more extras fitted as standard. Dearest of the lot, as always, is the Vanden Plas which has gone from $42,940 to $52,500.

Naturally, the latest government restriction on lease tax relief, announced at the budget, is going to affect sales to a great degree, but the Leyland people seem to feel that this will be only a temporary set back.

One thing the government move is likely to have achieved however, is to make buyers more selective on a basis of driving excellence and vehicle equipment, more than making the choice one of financial investment!

This being the case, of the reduced market for these $30,000 plus motor cars, Jaguar should take a fair share.

Lovers of fine motor cars will enjoy the solid "clunk" as the doors close, encapsulating the occupants away from all wordly cares in an environment which is climate controllled through the air conditioning system.

The facia layout is familiar to all those who have driven previous models, and the slim rimmed leather covered steering wheel is still there, albeit with a revised centre boss and spoke.

A feature of the front seat area we like is the economical use of space which gives the impression of a cockpit. It adds to a feeling of at-oneness with the car when driving, and with all the controls so readily to hand, little or no effort is required of the driver in operating the car.

The seats are adjustable fore and aft and in respect of backrest angle, but there's no height adjustment. The length of the steering column is adjustable too through a gnurled ring just behind the steering wheel itself.

Oddly enough, the indicator switch is located in the left hand steering column mounted stalk, obviously designed for the American market and remaining unchanged for right hand drive countries. The right hand stalk looks after the windscreen wipers and washers, as we said before, including an intermittent mode for the first time.

Push switches on the centre vertical portion of the console look after the fuel tank selection from the left or right 45.5 litre fuel tanks mounted either side of the boot. They register on the fuel gauge individually.

MOTOR MANUAL
ACTION ANALYSIS
Jaguar XJ6 Series III 4.2

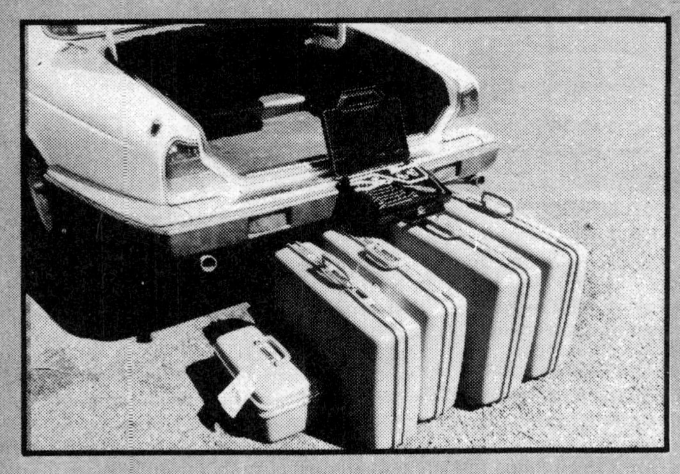

MODEL............ Jaguar "XJ6" 4.2 Series III
MANUFACTURER............ British Leyland
BODY TYPE/SEATS...... Four Door Sedan/5
PRICE -
Basic........................ $38,500
As Tested.................... $38,500

ENGINE:
Location.................................. Front
No. of cylinders........................ Six in line
Capacity (cc)............................ 4235cc
Bore and stroke (mm)........... 92.07 x 106
Block................................. Cast iron
Head................................. Aluminium
Valve gear........ Twin overhead camshafts
Induction Lucas Bosch electronic fuel injection
Compression ratio............................ 8.7:1
Max. power (kW/bhp)
152.9/186.9 at 5000 rpm
Max. torque (Nm/ft lbs)
314/231.7 at 1500 rpm

TRANSMISSION:
Type .... Three speed automatic — drive
Shift location........... Centre console T-bar
Ratios —
First.. 2.400
Second...................................... 1.460
Third.. 1.100
Final drive.................................. 3.07

BODY/CHASSIS:
Construction.............................. Unitary
Panel material.............................. Steel

SUSPENSION:
Front........ Independent double wishbones Coil springs swaybar
Rear .... independent lower transverse wishbones twin coil springs radius arms

STEERING:
Type ...... Rack and pionion - power assisted

BRAKES:
Type................ Hydraulic power assisted
Front........................... Ventilated disc
Rear..................................... Solid disc

WHEELS AND TYRES:
Wheel type........... Pressed steel ventilated
Diameter/rim width (in)................... 15"/6"
Tyre make/type...... Dunlop SP Sport radial
Dimensions............................ ER70 VR15

DIMENSIONS:
Weight (kg)................................ 1830
Length (mm)............................... 4959
Width.. 1770
Height....................................... 1377
Wheelbase.................................. 2865

Front track.................................. 1480
Rear track,.................................. 1495
Fuel tank capacity (litres)...... 2 x 45.5 litres

PERFORMANCE:
Speedometer error (km/h)

Indicated	60	80	100
Actual	57	76	95

Maximum speeds in gears (km/h)
First... 85
Second... 140
Third............................ 180 approx.

Acceleration from rest to —
60 km/h.......................... 5.1 secs.
80 km/h.......................... 7.3 secs
100 km/h........................ 10.3 secs.
120 km/h........................ 13.9 secs.

Acceleration from 60 km/h to —
100 km/h.......................... 6.8 secs.

Standing start to 400 metres —
Elapsed time..................... 18.05 secs.
Terminal speed (km/h)...................... 133

Braking —
110 km/h to rest (average)...... 50 metres

Fuel consumption (litres per 100 km/mpg)
Driven hard.................................. 18/15
Driven normally....................... 16.1/17.6

★ Performance figures recorded using SILICONIX ET 100 digital stop watches from Smiths Industries P/L, Technical Sales Division, 132 Bank Street, Melb. 3205

Cases used are Sampsonite "Saturn" supplied by Namco Industries (Vic.), Princes Hwy, Harrisfield.

Well formed Connolly leather seats feature throughout, those at the rear easily accommodating three adults if the need arises. There is ducting from the air conditioner into the rear passenger area too of course.

Electric windows are fitted all round, there being a centre armrest mounted switch bank under the control of the front seat passengers, with the rear passengers having a duplicate set for their windows, in the centre console ahead of them.

Without a doubt, the most frustrating thing about the Series 111 is its front seat belt mechanism. Time and time again the passenger found it impossible to unlock the belt in order to draw out sufficient length to fasten it. The same happened with the driver's seat belt, and quite often it was necessary to drive for a while without the belt on until it decided whether or not it was eventually going to co-operate!

If the car was parked on any sort of slope, whether fore an aft or to the side, again the belts simply refused to be drawn from out of their hiding place in the B pillar.

We're not sure whether it was a problem just on our test car, but we have found it before on a few other Series 111 models we've driven.

It is something that requires VERY urgent attention from Leyland as it doesn't take a genius to figure out just how serious such a fault can be.

The designers have considered interior carrying space quite well, with a useful tray fitted on the facia for odds and ends that might be needed in a hurry. There's also the usual lockable glove box in the carefully matched walnut dash.

At the rear there are pockets in the front seat backrests as well as the wide parcel tray under the rear window.

Boot space is not what you'd call generous, but it's sufficient for carefully selected luggage, and it is uniform in shape. A very fine tool kit comes with the car, nicely presented in a case complete with carrying handle.

Somehow, despite all the refinement designed into the car, those Jaguar guys have retained the pure adrenalin pumping feeling of being alive. You can feel it through the steering wheel, and you can feel it in the way the tyres bite on the road surface. Under brakes too there's every feature that one would expect from such a throughbred.

Simply crusing at 100 km/h isn't what its all about. Indeed, so fine are the ride characterisitc, and suppression of noise both from the road and the engine, that the speedometre becomes a very necessary instrument. Without it, an actual 130 or 140 km/h feels like 90!

Around town the feeling of calm actually makes driving almost a pleasure. The steering response allows the car to be pointed around the peasants with a minimum of fuss, and the sure footed braking gives instantaneous response.

Interior silence is best up to 130 km/h or so. Only after that with the engine starting to wind out towards the 5000 rpm red line does the engine noise start to percolate through. But at no time does it intrude in an offensive manner.

The series III and Series II compared; dimensions are similar but you'll note the Series Three is all new above the waistline.

Series III has neater, more agressive front end.

It's easy to recognise the new car from behind, it's the one with the sculptured tail lights.

Automatic shifts by the Borg Warner 66 transmission are hardly discernable. We rather think the change is not quite as good as that on the Mercedes three speed unit, but it's as near as makes no difference.

Because of relatively high gearing, initial acceleration is not in the sparkling category, but as soon as the engine starts to rev a little, the speed builds up quickly. Even so, a zero to 60 km/h figure of 5.1 seconds is pretty good, but not as outstanding as the 100 km/h figure of 10.3 seconds. For a car weighing over 1800 kilograms powered by 152 kW, that's something else.

In general, acceleration is best left to the automatic gearbox rather than using the T bar manually. That, if nothing else, proves that the ratios were well chosen.

When it comes to fuel consumption, our normal driving figure of 16.1 1/100 kms was not outstanding. This was achieved on a mixture of city, boulevard and freeway driving and is, we believe, pretty representative of what the average driver will enjoy.

On a long country trip we can accept something around 13 l/100 kms but use of the throttle will certainly see it drop away to the 18 1/100 km mark without too much trouble. Still that's better than the V12 XJ we drove some time ago which was hard pressed to do better than 28 1/100 kms! A later V12 tsted a short while prior to the introduction of the Series 111 range was good for 20.2 1/100 kms when driven on the same course as this latest car. The V12 also tends to be between one and two seconds faster on acceleration as might be supposed.

Handling and road holding are exactly what you would expect from a car with such a heritage. It's not a sports car, but a vivacious driver can do some pretty neat things with it, and there appear to be no vices.

There's a slight touch of understeer in there, but it is hardly noticeable and the vehicle can be cornered at very high speeds without it wanting to take over.

Under the conditions of our emergency stop from 110 km/h the Series 111 came to rest remarkably quickly, the length of road taken increasing only slightly over a concentrated series of six such stops. Initially the front right hand wheel tended to lock upsetting stability marginally, but then the front left also had a hiccough at another time, indicating that it was more the driver than the car that was at fault in not feeling the retardation out properly.

Whether Jaguar's reputation for lacking reliability is warranted or not, Leyland Australia's unique "Mastercare" plan must provide sufficient confidence for customers to take the plunge.

Each purchaser of any of the Jaguar or Daimler range gets a Mastercare kit which includes a personal identity card to obtain the service. This entitles the owner to a 24 month or 40,000 kilometre (whichever comes first) warranty, as well as free maintenance service for up to 40,000 kilometres or THREE years (whichever comes first). The owners pays only for the materials that are used.

Best of all, the Mastercare plan applies to all subsequent owners of the car until the time or distance limit is reached.

Mastercare is an Australian deal, peculiar to this country, and it must be quite an attraction for prospective customers investing such large amounts of money in a car of this type.

It has taken us a surprisingly long time to get rid of what was once a less than respectful view of Jaguars in Australia. With the Series 111 however, the quality of build is much improved. In addition the increased refinement makes it a very desirable car indeed. With Mastercare backing all this up, we rather imagine there will be a lot more such cars around, no matter how much Mr Howard wants to pay lip service to the ideal of bringing all cars down to the lowest common denominator! ☐

ELECTRONICALLY-FUEL-INJECTED

Jaguar 4.2

Does 0-60 in 9.6 secs, 122 mph and gets from 14 to 27.7 mpg.

ELECTRONIC fuel injection — along with styling/interior refinements — mark the newest Jaguars to come from Jaguar Rover Triumph Ltd. Previously limited to the larger 5.3-litre V12s, the new system is now standard on the 4.2-litre six, used in the XJ6 Series III. The result, says Jaguar Cars, is improved efficiency, economy, flexibility and an increase in maximum power of 30 bhp.

Factory performance figures show that with automatic and 3.31 axle, the 4.2 will do 0-60 mph in 9.6 secs., 122 mph and get fuel economy of 14 mpg in urban driving, 21.9 at 120 kph (73 mph) and 27.7 at 90 kph (55 mph).

A minor change in the roofline — to give more glass area along with more headroom for rear seat passengers — plus black wraparound bumpers and a new tail light cluster, are the only body changes.

A major change in the interior are the front seats, which now have adjustable lumbar support. Other modifications include: thick pile carpets, concealed front inertia seat belts and revised headlining.

Other mechanical changes include two options: cruise control and a 5-speed manual gearbox in place of the former 4-speed with overdrive. The larger XJ12 5.3 engine will continue to be fitted with standard automatic.

Jaguar hopes these changes will enable them to produce more than the 26,500 in 1978. Of those, the US took nearly 5000, a rise of 10%, maintaining its position as No. 1 overseas market.

Auto TEST Jaguar XJ6 4.2 Automatic

Still distilled excellence

Jaguar XJ6 4.2 Automatic
The XJ6 first appeared in 1968 in 2.8 and 4.2-litre form as a replacement for the much loved 3.8-litre Mk II saloon. In April 1975 the XJ 3.4 superseded the 2.8-litre as Jaguar's "economy" model. This normally aspirated engine has run alongside the 4.2-litre six and 5.3-litre V12 saloons (introduced September 1973) ever since. Lucas/Bosch/Bendix D Jetronic fuel injection was first used on the V12 XJ Coupé, XJS, and V12 saloons in 1975. Lucas/Bosch L-Jetronic became standard on the more powerful 4.2-litre sixes with the introduction in April this year of the much improved and restyled Series III XJ6.

PRODUCED BY:
Jaguar Rover Triumph Ltd.,
Browns Lane,
Allesley,
Coventry CV5 9DR

MAINTAINING the standard of excellence and value for money set by the XJ6 on its first appearance cannot have been easy for Jaguar, and to have done it through detail improvement of what is now a 12-year-old design is all the more remarkable.

Its styling was individual and right then, and although opinions vary about the latest facelift most think it still is. Even now sheet metal changes have not been for change's sake. A higher roof line has been adopted at the rear to give back seat occupants more headroom. Deeper windows give more glass area and there are now no front quarter lights. The revised body has considerably more tumble home and therefore a narrower roof and slightly smaller frontal area. Other features are flush fitting door handles, a larger rear light cluster, deep rubber inserted bumpers with inset rear fog lights and indicators at the front. The sidelights are now built into the outer headlamp units. Above all the updating is discreet, and, say Jaguar, results in a marginal improvement in aerodynamic efficiency. A Cd figure of 0.412 is quoted.

Detail changes to the standard specification are dealt with later, but it should be mentioned that the test car was fitted with the optional air conditioning, AE Econocruise control, inertia reel rear seat belts, electric remote control door mirrors and electric seat height adjustment (adjustable lumbar support is standard) but not the optional electric sliding roof. These and other extras raise the price to over £17,300.

Suspension settings are unchanged, but Jaguar have taken the opportunity to alter the degree of power steering assistance slightly, by modifications to the valving and ram actuating torsion bar. Interestingly the test car was fitted with, recently approved by Jaguar, Pirelli P5 205/15VR tyres.

By today's standards the Jaguar is heavy, but at 34.6 cwt (distribution 54.5/45.5 front to rear) no more so than its predecessor. Where Jaguar have made a really significant advance is in engine efficiency. The adoption of Lucas/Bosch L-Jetronic electronic fuel injection, a higher compression ratio (up from 7.8 to 8.7 to 1), bigger inlet valves, and earlier inlet camshaft timing, means that the power output from Jaguar's seemingly timeless — certainly delightful — long stroke 4.2-litre DOHC straight six goes up from 180 bhp (DIN) at 4,500 rpm to 205 bhp at 5,000 rpm. Peak torque rises from 232 lb. ft. at 3,000 rpm to 236 lb. ft. at a higher 3,750 rpm, yet it is still an extraordinarily flat torque curve, with over 210 lb. ft. being produced from as low as 1,600 rpm.

In the new car Borg Warner's model 66 transmission supersedes the model 65. Both have the same internal ratios, but the new box has a number of detail improvements to give it a higher torque capacity. With a fatter

The new body is characterised by deeper glass all round, a higher roof line at the rear, recessed door handles, and deep rubber inserted bumpers incorporating fog lights at the rear and indicators at the front. The rear light clusters have also been enlarged.
The optional headlamp wipers work very effectively, but on the outer dipped beam lamps only

wer curve and a higher power ak Jaguar have raised the final ve ratio from 3.31 to 1 (the d model was markedly dergeared) to 3.07 to 1, thus sing overall gearing from .9 per 1,000 rpm to the XJ 2's 24.7 mph per 1,000 n.

erformance
amatically better

As we have come to expect of y L-Jetronic equipped engine, ·ly morning starts are virtually tant; also the warm up driving riod is utterly hesitation- e. Jaguar's attention to tting a quick warm means the water tem- rature gauge registers nin a quarter of a mile, but re pleasingly for the occu- nts, the heater begins to take ect even before that.

How does it go? Only a glance the figures (earlier car's ones brackets) is needed to realise w much performance has en transformed, particularly at top end. We should mention t if left to its own devices the arbox will change from 1 to 2 4,300 rpm, but from 2 to D at ideal 5,000 rpm peak power int. For obtaining the best ssible straight line accelera- n figures Jaguar recom- nded that we hold first nually to the red line, then the lever through to D and let gearbox do the rest. So doing ed 0.8 sec to 60 mph and sec to 110 mph during the celeration runs. Initial eleration from rest was still te leisurely. Zero to 30 took 3 sec (3.9 sec), 60 mph came in 10.0 sec (10.6 sec), reafter the new car draws ay ever increasingly, 80 mph ing 16.9 sec (18.5 sec), 100 h 29.4 sec (34.1 sec) and 0 mph 38.5 sec — a remark- e 13.3 sec quicker than be- e. Equally dramatic improve- nts are seen in the gears with 90-110 mph in top time tually halved to 16.2 sec eviously a lazy 32.3 sec), 30- mph in 1 took 3.3 sec (4.3) and 60-80 mph in 2 6.7 tead of 8.5 sec. Whereas the lier car was overreving 600 n to achieve its maximum of 7 mph, the Series III only ayed slightly over peak power 5,000 rpm and into the red ctor in pulling a mean 127 h (5,150 rpm) and a best one y speed of 130 mph (5,250 n), so can be considered just out ideally geared. Inciden- lly, all performance figures were taken with the heater on — in which mode the air conditioning pump is intermittently working which imposes an extra load on the engine. Good weather conditions prevailed during the maximum speed runs with a light wind running in the car's direction. No stability problems were evident; however, later high speed testing under light cross wind conditions did cause the car to require constant, if slight, steering corrections to keep it running straight.

Left to its own devices the Borg Warner Transmission shifts smoothly and, as we have said, at 4,300 and 5,000 rpm, an indicated 43 and 83 mph. If held manually maxima in the gears are in fact 51 and 83 mph. Maximum speeds at which kickdown is available are, however, much lower; not above 30 mph from 2 to 1, or over 67 mph from D to 2. Part throttle kickdown only occurs well below these speeds. This sometimes makes the car seem lacking in go, particularly when immediate acceleration response is required just above those speeds. The answer is, of course, to select gears manually. As ever, this brings us to the potentially dangerous Jaguar selector arrangement, which remains unchanged even after constant criticism from *Autocar* and others. Firstly, there is no detent to stop the driver from pulling that delightfully slender T handle from D straight back into 1 at too high a road speed (fortunately an inhibitor in the gearbox prevents engagement of either gear). More inexplicable is the provision of a detent between 2 and D yet none between D and neutral. Overcoming the 2 to D stop requires the driver to use sideways pressure before moving the selector forward but it is then free to move through to neutral uninhibited with potentially dire consequences for the engine, or even the car if in mid full throttle overtaking manoeuvre. Worse still, if the side pressure is not relaxed the selector can be pushed straight through to reverse. The former mistake would be all too easy for the unfamiliar, or nervous driver to make, and the latter would certainly send the car out of control, and does not bear thinking about. All this is a pity because in almost every other respect the new XJ is a driver's delight.

Power is available from very low down in the rev range to take one wafting effortlessly past a

line of traffic. The car responds with smooth obedience. Its slow revving engine never feels or sounds stressed, even when pressed hard. Throttle action is beautifully silky and progressive. A natural cruising gait is in the 90-110 mph bracket where it feels utterly relaxed and stable.

As mentioned the test car was fitted with cruise control. After switching in via a rocker switch situated conveniently behind the selector gate, the chosen cruising speed is selected by pressing a button on the end of the left-hand column stalk. The car will return to the chosen speed after the driver accelerates unless, of course, he selects a higher speed or cancels by touching the brakes or switching it off. We were particularly impressed by the gentle way the AE Econocruise kept the desired speed, unlike some others which seem to use coarse throttle movements, thus creating annoying surges in acceleration.

Economy
Much improved

Accelerating a high performance car as heavy as the Jaguar will never be an economical business. Yet an overall consumption figure of 16.8 mpg (15.0 mpg last time) together with the very real performance gains shows how much Jaguar have increased the car's efficiency. During a very lengthy and enjoyable test period, consumption varied from 13-14 mpg (testing and commuting in snarled up London) to nearly 20 mpg obtained on one long and gentle cross country trip. With consideration owners should easily average between 18 and 20 mpg, giving them a comfortable range of 300 miles on the XJ6's split 22 gallon tankage. At least one tester fell into the habit of running one 11 gallon tank dry before changing over to the other, via the familiar centre console mounted push button switch, an almost instantaneous pick up making this quite a reasonable procedure. The wing top fuel fillers accept full pump flow without problem until one tries to squeeze in the last gallon or so, which takes several frustrating minutes. The test car suffered a slightly high, by today's standards, oil consumption of 800 miles per half litre.

Noise
Worse for P5s but still among the quietest

Better handling response is claimed for the new Pirelli P5s; however they undoubtedly generate more roar and rumble than the optional, previously fitted, and exceptionally quiet-running Dunlops. This is particularly noticeable over ridged concrete or coarse and uneven surfaces. Cats' eyes are noticed more and running over repaired roads, drains or potholes slowly produces a quite pronounced (by Jaguar standards) rumble and thump through the bodyshell. There also appeared to be a slightly higher level of wind noise above 80 mph than we remember on previous XJ6 test cars. That said, wind noise increased only marginally at much higher speeds, say when running between 100 and 120 mph, but as before one is distantly rather than acutely aware of this, and it is the *only* real noise one has to suffer on the motorway. The Jaguar XJ6 remains a supremely quiet and relaxing car to ride in. At tickover (600 rpm) the engine is sensed beating rather than heard. At rest the peaceful environment inside is disturbed only by the clock ticking and the little hisses, whines and distant fan noise produced by the automatic heating and ventilation system. It is beautifully quiet at town speeds and cruising in the 40-50 mph range the most evident sound may be some muted tyre roar. Even when the engine is worked its deep and willing note remains muted yet evident enough for the enthusiast to relish. When you switch off the electric aerial retracts almost inaudibly after a 10 sec delay. This prevents the aerial "hunting" if you stall.

Road Behaviour
Impeccable

As we have mentioned, Jaguar have at last taken the opportunity on the new XJ6 to slightly reduce the amount of power steering assistance. The degree of heavying up is small and

An impressive engine bay full (right). Access to daily service items is good

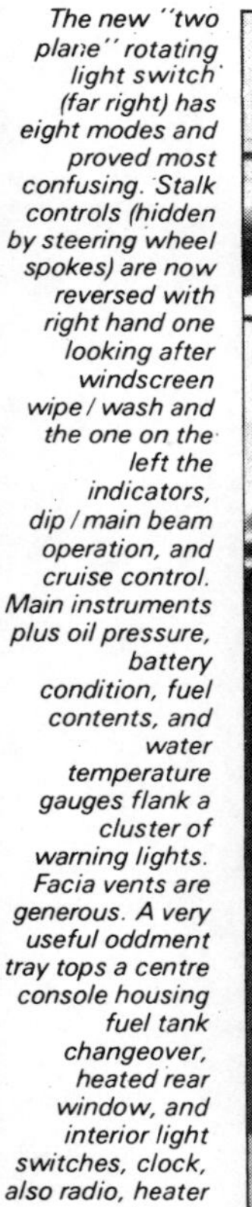

Undoing the screw type oil filler cap (above) is a finger scorching exercise if the engine is hot

could usefully be taken further, but it is noticeable and provides a very welcome increase in feel. Gearing at 3.3 turns from lock to lock remains a shade on the low side especially when one considers the car's very poor 41ft turning circle, which can lead occasionally to an embarrassing wait for traffic to exit a turning into which one wishes to turn. Also the rounded bonnet and slightly bulging sides of the car make the Jaguar less easy than say the more squared off Granada shape to thread through narrow gaps with confidence.

Jaguar's combination of ride, handling and roadholding remains a superb one. Without conducting back-to-back Pirelli vs Dunlop tests we cannot say for certain whether the gains in handling response we felt due to P5s are actually worth the increased road noise, and, incidentally, the slightly extra expense when they are replaced, but even with P5s poor surfaces and particularly undulating ones are soaked up in a remarkable way, though we did notice that the poorly maintained country road taken at speed sometimes sent a tremor through the bodyshell. Generally the occupants get an extraordinarily serene ride. Some bumps are felt, but the suspension is hardly, if ever, caught out. The wheels stay glued to the road and the driver never loses his sense of control.

As one would expect in a heavy front-engined car, natural straight line stability is excellent. Cross winds, however, do push the car around noticeably, though not at all unnervingly. At very high, therefore possibly somewhat academic speeds, the steering begins to feel overlight and a shade imprecise, particularly when the road surface is wet. Then delicate fingertip rather than hand movements are best to correct wander.

The cornering and traction grip of the Pirelli 205/70 VR tyres is excellent in the dry or wet. In normal driving handling balance is neutral. In low gear on wet roads there is enough instant driving torque on tap to spin the offside rear wheel (no limited slip is fitted) or to provoke the car into a rear end slide. In the event, correction is quick, especially so for a heavy car. As we have suggested, the car's dry road cornering limit is extremely high; more to the point, on limit handling is entirely quirk free thanks mainly to a proper low camber change non semi-trailing arm rear suspension.

As the limit is approached roll builds up but never to the point of making the driver feel insecure, and less so than one might imagine considering the car's soft, but superbly controlled, straight line ride. The steering has feel as the front tyres be to slide. Under power, m understeer prevails while t Jaguar rear end stays resolu glued to the road. Lifting off this point will cause the rear e to move out but very docilely a to a degree that hardly requi correction. Overall it is still XJ6's ability to corner at di ming speeds unobtrusively, to be upset by irregular surfa and to cruise safely and qui at very high speeds that cha the driver.

Further proof of the Jagua thoroughbred nature w brakes that easily shrugged the stiffer *Autocar* fade test (consecutive 0.5g stops fr 82mph). After the initial sp sensitiveness had worn pedal pressures rose slightly, then remained absolutely c sistant between the fourth a last stop. The only sign of punishment they were tak was some roughness and sm produced during the last f stops. Pedal pressures are lig yet there is no suggestion of brakes "biting back" initially so often happens nowada leading to criticisms of overs sitiveness. This seems d mainly to Jaguar's ability to g very progressive pedal fe Pedal pressures varied in a v linear manner from a 10 caress sufficient to prod 0.25g deceleration — the ge speed check — to a mild 4 shove which produced a q outstanding best stop of o 1.0g. In dry test conditi further pedal effort sim locked the front wheels, s thing we later discovered the

The new "two plane" rotating light switch (far right) has eight modes and proved most confusing. Stalk controls (hidden by steering wheel spokes) are now reversed with right hand one looking after windscreen wipe / wash and the one on the left the indicators, dip / main beam operation, and cruise control. Main instruments plus oil pressure, battery condition, fuel contents, and water temperature gauges flank a cluster of warning lights. Facia vents are generous. A very useful oddment tray tops a centre console housing fuel tank changeover, heated rear window, and interior light switches, clock, also radio, heater controls

was rather too prone to do in wet.

With a big heave the pul- umbrella handbrake just held Jaguar in MIRA's 1-in-3 slope and managed a v creditable 0.37g retardation the flat.

Behind the Wheel

Sometimes confusing

The seats have plenty of re ward movement, and a n welcome three position adj able lumbar support. Elect seat height adjustment (wh fitted) tips the driver's seat ward as it rises so it can be u to iron out the coarseness of ratchet type back rest adj ment. or alternatively the c umn can be adjusted for plun Although not looking obviou comfortable we found them under all but the hardest of c nering when one is apt to sl about rather on their sh leather facing (cloth is a no-c option).

Interior finish is high quality and traditional with a wood veneered dash and leather upholstery. The rear seat passengers have a ventilation outlet in the back of the centre console, also window switches and door speakers.
The large lever in the seat side operates ratchet type backrest adjustment and a rocker switch on the seat cushion corner, the seat height adjustment

All round visibility seems little ected by the thicker rear pil- s and remains good, except in t weather when the symme- cally pivoted wipers (now anged to park conventionally) ve both corners of the screen wiped.

For such an elaborate car the ntrols, instruments and war- ng lights are refreshingly nple. The revcounter and un- ually accurate speedometer ntain heated rear window and lb failure warning lamps. If latter remains on for more n 30 sec after any lights have en switched or the handbrake d footbrake are operated ether, bulb failure has occur- l.

Existing Jaguar drivers will d the same centre console using; clock, heated rear ndow, fuel tank change over, p reading and interior light itches (a delay keeps this on 10 secs after the door is osed). Underneath are the dio and heater controls and ectric window switches. The ndows incidently work toler- y fast.

Stalk controls have now been ught into line with European tate, the right hand one now oking after windscreen sh and two speed plus long last automatic inter- tent operation as well as flick pe. Operating the washers th the headlights on also ngs the headlamp wash and ers (these only work on the ter dipped beam lights) into eration.

Apart from housing the speed ntrol button, the left hand lk operates indicators, and (by lling towards the driver) a rmal headlamp flash when the in lights are switched off, but driver has to methodically and un-dip the lights to give rning with the headlights in e.

A minor complaint is that Jaguar did not take the opportunity to redesign the stalk controls to move with the steering column. The taller driver who likes the column adjusted for an arms bent stance finds the stalks awkwardly at full finger stretch. Terribly confusing is the new "two plane" rotating light switch. Pulled out (the normal plane) and turned clockwise it switches in sidelights, headlights, and finally the rear fog lights at the same time dipping the headlights if on full beam. If the knob is then pushed in the headlights are dipped automatically (this facility is blanked off where no front fog lights are fitted) and the front fog lights come on. Turned one click clockwise the headlights go out, leaving the fogs on. Two clicks anti-clockwise brings all the lights into operation and, finally, a third switches out all but the front fogs and sidelights. Eight lighting modes in one switch are far too many for any driver to remember in the heat of the moment, and one inevitably asks the question, are they all necessary?

Changes to the central locking mean that the system works more conventionally, albeit still noisily by solenoids via the door key rather than by switch in the centre console. The boot then stays locked until unlocked separately with a key. Thus valuables can be left secure while the car is driven by someone other than the owner.

The excellent Delanair integrated air blending heater and air conditioning system was fitted to the test car. Temperatures between 65 and 85 deg. F. can be selected and then the desired fan mode chosen from Off-Lo-Auto-Hi-Def. It is noisy when working hard to defrost the screen or for that matter, when pulling the temperature of the car down after it has been cooking in the sun, but once the temperature is under control it is maintained quietly. What the driver does not have at the moment is the ability to alter the temperature split between feet and face. Indeed any extra fresh air ventilation (via pull handles under each side of the facia) is at feet instead of at face level where it is most needed. We understand changes will be incorporated soon to give the occupants split level control.

Living with the XJ6

For the rear passengers there is ventilation at feet level and one rotating upper level outlet in the rear of the transmission console. Most found the rear seats comfortable. Tall souls had adequate headroom, but found their shins banging on the frame tubes that run across the bottom of the front seat backs.

Yet another new feature is the provision of an electric aerial that can be "tuned" in length via a rocker switch to get, say Jaguar, better VHF reception, particularly in America. As it was, reception and sound reproduction from the Philips A860 radio-cum-cassette player were excellent.

A minor inconvenience is that the increased tumble home means that a few drips of water fall on to the seats when the doors are first opened after a stay out in the rain. Oddment space is fairly limited in volume, but there are a decent number of compartments; the glovebox, arm rest locker and four door pockets.

The boot is shallow, but makes up for this in length. It remains a difficult shape for hard suitcases. Well planned use of soft luggage would increase the effective volume somewhat. The tools and first aid kit are clipped awkwardly at the forward end of the boot, the jack and wheel-brace to the side, while the spare wheel is housed under the boot floor.

The spare wheel is housed under boot floor. Boot space is cluttered by tools and first aid kit (forward end on left) also jack and warning triangle (clipped to left hand side)

Annoyingly, the engine bay is not illuminated, making location of a minor night time fault a chore. Fan or alternator belt replacement looks awkward, otherwise daily service items are easy to locate and the classically beautiful Jaguar engine always a pleasure to look at.

Mention finally must be made that Jaguar still requires service attention at 3,000 mile intervals which by today's standards is far too frequent. Yet this, and the few other failings the series III XJ6 has — particularly that selector arrangement — are far outweighed by the its extraordinary qualities.

The Jaguar XJ Series III Range

As before there are three: the carburettor XJ6 3.4 costing £13,259, the fuel injected XJ 4.2 at £14,609 and the incomparable XJ12 5.3 priced at £17,627. Automatic or five speed manual transmission are available on six cylinder XJs at no difference in cost, but the XJ12 5.3 can only be had with the GM400 automatic transmission.

HOW THE JAGUAR XJ6 4.2 (A) PERFORMS

Figures taken at 2,783 miles by our own staff at the Motor Industry Research Association proving ground at Nuneaton.

TEST CONDITIONS:
Wind: 10-15 mph
Temperature: 10 deg C (50 deg F)
Barometer: 29.3in. Hg (993 mbar)
Humidity: 80 per cent
Surface: dry asphalt and concrete
Test distance: 2,893 miles

MAXIMUM SPEEDS

Gear	mph	kph	rpm
Top (mean)	**127**	204	5,150
(best)	**130**	209	5,250
2nd	**85**	137	5,000
1st	**51**	82	5,000

ACCELERATION

FROM REST

True mph	Time (sec)	Speedo mph
30	**3.8**	30
40	**5.6**	40
50	**7.4**	50
60	**10.0**	60
70	**13.3**	70
80	**16.9**	80
90	**21.9**	91
100	**29.4**	101
110	**38.5**	111

Standing ¼-mile: 17.4 sec, 82 mph
Standing km: 31.6 sec, 103 mph

IN EACH GEAR

mph	Top	2nd	1st
10-30	—	—	2.9
20-40	—	—	3.3
30-50	—	5.3	3.3
40-60	—	5.5	—
50-70	—	5.9	—
60-80	—	6.7	—
70-90	10.7	—	—
80-100	12.5	—	—
90-110	16.2	—	—

FUEL CONSUMPTION

Overall mpg:
16.8 (16.9 litres/100km)

Constant speed:
Autocar constant speed fuel metering equipment incompatible with fuel injection.

Autocar formula: Hard 15.1 mpg
Driving and conditions: Average 18.5 mpg
Gentle 21.8 mpg

Grade of fuel: Premium, 4-star (97 RM)
Fuel tank: 20 Imp galls (91 litres)
Mileage recorder: 1.3 per cent long.

Official fuel consumption figures
(ECE laboratory test conditions; not necessarily related to Autocar figures)
Urban cycle: 14.5 mpg
Steady 56 mph: 28.2 mpg
Steady 75 mph: 23.7 mpg

OIL CONSUMPTION

(SAE 20/50) 800 miles/pint.

BRAKING

Fade *(from 83 mph in neutral)*
Pedal load for 0.5g stops in lb

	start/end		start/end
1	20/12	6	28/32
2	20/24	7	28/32
3	24/24	8	28/32
4	28/24	9	28/32
5	28/28	10	28/32

Response (from 30 mph in neutral)

Load	g	Distance
10lb	0.25	120ft
20lb	0.51	59ft
30lb	0.70	43ft
40lb	1.01	30ft
Handbrake	0.37	81ft

Max. gradient: 1 in 3.

WEIGHT

Kerb, 34.6cwt/3,875lb/1,760kg
(Distribution F/R, 54.5/45.5)
Test, 38.0 cwt/4,255lb/1,932kg
Max. payload 900lb/409kg.

DIMENSIONS

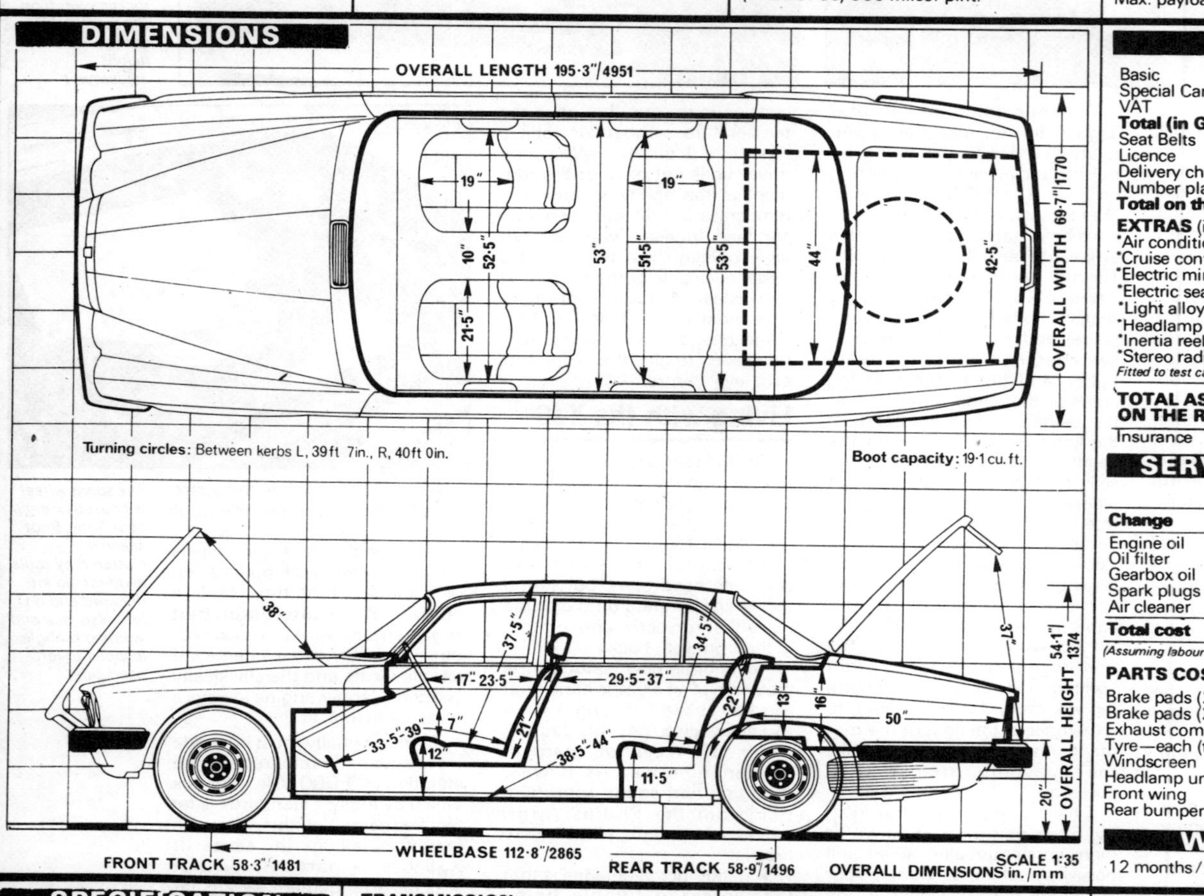

PRICES

Basic	£11,726.0
Special Car Tax	£977.1
VAT	£1,905.4
Total (in GB)	**£14,608.6**
Seat Belts	In
Licence	£50.0
Delivery charge (London)	£50.0
Number plates	£15.0
Total on the Road	**£14,723.6**

EXTRAS (inc. VAT)

*Air conditioning	£1,109.4
*Cruise control	£287.8
*Electric mirrors	£103.6
*Electric seat adjustment	£168.5
*Light alloy wheels	£459.6
*Headlamp wash/wipe	£184.2
*Inertia reel rear seat belts	£66.4
*Stereo radio/cassette player	£210.2

Fitted to test car

TOTAL AS TESTED ON THE ROAD	**£17,313.**
Insurance	Group

SERVICE & PARTS

Change	Interval 3,000	6,000	12,000
Engine oil	Check	Change	Change
Oil filter	—	Change	Change
Gearbox oil	Check	Check	Check
Spark plugs	—	Change	Change
Air cleaner	—	—	Change
Total cost	**£18.40**	**£64.78**	**£78.**

(Assuming labour at £8.00/hour)

PARTS COST *(including VAT)*

Brake pads (2 wheels)—front	£23.3
Brake pads (2 wheels)—rear	£11.7
Exhaust complete	£231.1
Tyre—each (typical)	£106.4
Windscreen	£70.7
Headlamp unit	£20.9
Front wing	£110.4
Rear bumper	£68.0

WARRANTY

12 months/unlimited mileage.

SPECIFICATION

ENGINE

Head/block	Front, rear drive; Al. alloy/cast iron
Cylinders	6, in line
Main bearings	7
Cooling	Water
Fan	Viscous
Bore, mm (in.)	92.07 (3.62)
Stroke, mm (in.)	106.00 (4.17)
Capacity, cc (in³)	4,235 (258)
Valve gear	Dohc
Camshaft drive	Chain
Compression ratio	8.7-to-1
Ignition	Breakerless
Fuel injection	Lucas/Bosch L Jetronic
Max power	205 bhp (DIN) at 5,000 rpm
Max torque	236lb ft at 3,700 rpm

TRANSMISSION

Type: Borg Warner model 66 3-speed automatic

Gear	Ratio	mph/1,000 rpm
Top	1.00-2.00	24.7
Inter	1.45-2.90	17.1
Low	2.39-3.78	10.3

Assuming no torque converter slip.

Final drive gear: Hypoid bevel
Ratio: 3.07

SUSPENSION

Front—location	Double wishbones
springs	Coil
dampers	Telescopic
anti-roll bar	Yes
Rear—location	Lower wishbones and fixed length drive shafts
springs	Coil
dampers	Telescopic
anti-roll bar	No

STEERING

Type	Rack and pinion
Power assistance	Yes
Wheel diameter	15¾in.
Turns lock to lock	3.3

BRAKES

Circuits	Two, split front/rear
Front	11.2in. dia. disc
Rear	10.4in. dia. disc
Servo	Yes
Handbrake	Umbrella, on rear discs

WHEELS

Type	pressed steel
Rim width	6.0
Tyres—make	Pirelli P5
—type	Radial ply tubeless
—size	205/70 VR15
—pressures	F28, R26 psi (normal driving)

EQUIPMENT

Battery	12 V 66 Ah
Alternator	60 A
Headlamps	110/230 Watt to-
Reversing lamp	Standard
Hazard warning	Standard
Electric fuses	22
Screen wipers	2-speed plus intermittent
Screen washer	Electric
Interior heater	Air blending
Air conditioning	Extra
Interior trim	Leather seats, cloth headlining
Floor covering	Carpet
Jack	Screw scissor type
Jacking points	Two each side, under sills
Windscreen	Laminated
Underbody protection	Bitumastic underbody sealer plus wax injection

HOW THE JAGUAR XJ6 4.2 (A) COMPARES

Jaguar XJ6 4.2 Series III (A) £14,609

Front engine, rear drive

Capacity
4,235 c.c.

Power
205 bhp (DIN)
at 5,000 rpm

Weight
3,875 lb/1,760 kg

Autotest
22 December 1979

BMW 732i (A) £14,193

Front engine, rear drive

Capacity
3,210 c.c.

Power
197 bhp (DIN)
at 5,500 rpm

Weight
3,471 lb/1,574 kg

Autotest
17 November 1979
of manual 732i

Ford Granada Ghia 2.8i (A) £10,018

Front engine, rear drive

Capacity
2,792 c.c.

Power
60 bhp (DIN)
at 5,700 rpm

Weight
2,971 lb/1,350 kg

Autotest
8 February 1978
of manual 2.8iS

Mercedes-Benz 450 SEL (A) £19,161

Front engine, rear drive

Capacity
,520 c.c.

Power
25 bhp (DIN)
t 5,000 rpm

Weight
,904 lb/1,772 kg

Autotest
May 1974

Opel Senator (A) £11,365

Front engine, rear drive

Capacity
2,968 c.c.

Power
80 bhp (Din)
at 5,800 rpm

Weight
3,080 lb/1,400 kg

Autotest
1 November 1978

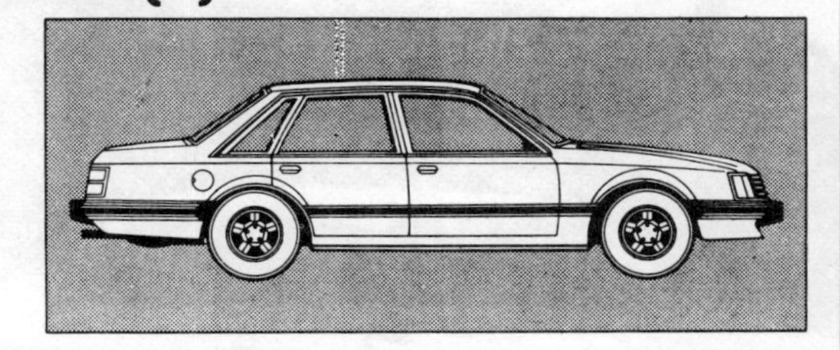

Peugeot 604 TI (A) £9,669

Front engine, rear drive

Capacity
2,664 c.c.

Power
136 bhp (DIN)
at 5,750 rpm

Weight
3,111 lb/1,412 kg

Autotest
22 November 1975
of manual 604 SL

MPH & MPG

Maximum speed (mph)

Mercedes-Benz 450 SEL (A)	134
BMW 732i	127
Jaguar XJ6 4.2 (A)	127
Opel Senator (A)	119
Ford Granada 2.8iS	117
Peugeot 604	113

Acceleration 0-60 (sec)

BMW 732i	8.0
Ford Granada 2.8iS	8.9
Mercedes-Benz 450 SEL (A)	9.1
Opel Senator (A)	9.2
Peugeot 604	9.4
Jaguar XJ6 4.2 (A)	10.0

Overall mpg

Ford Granada 2.8i Ghia	20.8
BMW 732i	19.8
Peugeot 604	19.6
Opel Senator (A)	18.6
Jaguar XJ6 4.2 (A)	16.8
Mercedes 450 SEL (A)	14.7

The truly rich might never consider much other than the Jaguar's really expensive opposition; perhaps a Bentley T2 or Rolls Royce Silver Shadow II (£36,652), a Bristol 603 S2 (£36,204), a Ferrari 400i (£31,809), or a de Tomaso Deauville (£24,418). The Maserati Quattroporte is available in lhd form for £28,900 and most obvious American, the Cadillac Fleetwood for £15,739.

We have chosen to compare the Jaguar with less exotic machinery three of which — the Granada, BMW and Peugeot — are manual (through lack of test data) and therefore likely to be around 10 per cent less economical than shown and marginally slower. The Mercedes is the most powerful, the heaviest, the fastest and predictably the least economical. Once on the move the Jaguar goes in the most pleasing, quiet (except under hard acceleration) and relaxed low revving way with much increased efficiency over its predecessor. The lighter and smaller engined Peugeot, Opel, Ford and very efficient "electronic" BMW go according to power outputs, the BMW particularly zestily. All are refined but all must be rated far less so than the XJ6.

ON THE ROAD

Although Pirelli P5s on the test Jaguar pushed up road noise over poor surfaces noticeably it still rates lower than any of our five. Jaguar/P5 noise and ride may now be on a par with or even marginally worse than a Rolls or Bristol, yet these cannot compete with the XJ6's combination of ride, handling and roadholding. Jaguar steering now has more feel for being slightly more heavily weighted. On balance though we still prefer the heavier and higher geared (2.75 turns) Mercedes system. In extremis there would be little to choose between the high cornering or predictable handling qualities of the Peugeot, Mercedes or Opel, with the Granada rating nearly as good but in Ghia form (there are stiffer S suspension packs available) more wallowy than the others.

Thanks to its low camber change rear suspension the XJ6 also corners at a very high limit — disarmingly fast for a big car — with little power on/off character change, unlike the BMW where thanks to its infamously compromising rear end, sharp rear end breakaway can be provoked by lifting the throttle sharply in mid corner.

Decent air blending heating and ventilation are to be found in the Jaguar, Ford, and BMW 732i (the latter is as complex to operate as the Jaguar and Ford are simple), Mercedes and Opel, though the last two are lower on flow. A water valve system lets the Peugeot down here, whereas the only thing that really dents the Jaguar's continuing top rating overall must be its poor lock — and that selector arrangement.

SIZE & SPACE

Legroom front/rear (in)

(seats fully back)	
Mercedes 450 SEL (A)	43/39
BMW 732i	42/40
Ford Granada 2.8i Ghia	41/40
Peugeot 604 SL	38/41
Opel Senator (A)	39/39
Jaguar XJ6 4.2 (A)	39/37

None of the group will stint the occupants for leg room front or rear, though the Jaguar's low rear seat and rail along the front seat back bottom make it less comfortable in the back than the BMW, the longest wheelbase Mercedes, or even the Granada. Biggest all round, the BMW has by far the largest boot, while the much more compact Senator has the smallest though not necessarily the least useful one. In sheer boot volume, the rest fall between these two (the Peugeot is second only to the BMW).

VERDICT

Jaguar have achieved a remarkable efficiency gain with the still comparatively heavy series III XJ6. It is faster, more economical, still has superb ride even on Pirellis (Dunlops are available), is extraordinarily quiet and handles for the most part like the true thoroughbred that it is.

Your choice will, of course, depend on what value you put on the Jaguar's quite unique qualities. In this company it must rate highest overall with the undoubtedly fast, tough, well finished but thirsty Mercedes and in comparison with a Rolls-Royce or Bristol — and it does compare in so many ways — the Jaguar is cheap. The BMW is let down by skittish handling particularly in the wet, while the Opel, Peugeot, and Granada could be grouped as lighter, more economical to run overall, and extremely competent cars.

The Jaguar remains a less compromising thoroughbred — in fact one of the few jewels the British motor industry has left to show the world.

Jaguar XJ6 Series III

Slick as the inside of Faye Dunaway's dressing gown.

• *The Jaguar XJ6 is the culmination of the efforts that have gone into SS and Jaguar cars since William Lyons first started putting special coachwork on small Austins and Standards. The philosophy then, was to produce a balance of elements not previously achieved in any fine motor car. The elements are style, performance, and luxury.*

Essential in all expensive cars, these three things are combined in a very happy manner in Jaguars; in such a way that one does not detract in the least from either of the others. The XJ is a complete car.

—Statement from Jaguar Rover Triumph

The Series III Jaguar XJ6—a refined and improved replacement for the old XJ6L—is one of the Western World's more delightful mechanical manifestations, and anyone who has more than $20,000 to spend on a new car should take a test drive in one before making his or her purchase decision. Unfortunately, American prejudice and conventional wisdom are stacked against British cars. This is partly because our national love-hate relationship with the country that mothered us three centuries ago has always been a prickly one (we'll probably never know how large a percentage of our population was actually pro-German at the outbreak of World War II), and partly because unreliable electrical components have done so much to undermine the reputation of British cars on American roads. A lot of this bad repute is myth—there are plenty of contented British car owners here in Columbia, the Gem of the Ocean—but it's a powerful myth, reinforced by just enough truth to keep it ever at the fronts of our minds.

This is a damned shame, because it will keep large numbers of Americans from going out and sampling an XJ6—surely one of the silkiest, most satisfying luxury cars available in this country. The XJ6 is as slick as the inside of Faye Dunaway's dressing gown. We could take that simile a step further, but Miss Dunaway's legal advisers would be on the phone in a flash. Suffice it to say that the inside of an XJ is as nice a place as any you can imagine. The XJ, either XJ6 or XJ12, is as rewarding a car as you're going to find on the public roads today—as refined as any Mercedes-Benz or BMW, but utterly different. That difference was demonstrated very effectively once in an old Robert Taylor movie: Robert Taylor played a crusader bragging about the raw hacking and hewing power of his great broadsword, when his Moorish adversary demonstrated the strength of his own weapon by tossing a silk scarf into the air to be sliced neatly in half as it wafted down over the slender Damascus blade . . . a Jaguar man if I ever saw one.

If possible, the XJ is both the Damascus blade *and* the silken scarf. It's smooth, effortless, and quiet—except for some wind roar where the windshield pillars and the outside mirrors come together—and it goes like a scalded dog. It is a truly luxurious sedan that should never, ever be driven at less than 80 miles per hour, especially on twisty, uneven two-lane pavement. If your first drive in an XJ is on a relatively smooth Interstate highway you'll love the interior of the Jag and gain the impression that it's about like any other fine car going down the road. If, on the other hand, your first XJ experience is on the above-mentioned twisty two-lane, you'll be stunned. The Jaguar tears across the nastiness just as quickly and safely as any big-buck German car; but it doesn't attack like the broadsword, it slips through like the Damascus blade. "Swift" is the word that comes to mind. You are not urged on with battle cries and exhortations; you are seduced.

What makes it such a seductive device? The basic specifications are such that it could have been produced in England, Germany, or Italy, but its personality sets it apart. I wouldn't automatically categorize it as an "English" car, but it certainly doesn't feel either German or Italian. Engineering tradition plays an important part. This six-cylinder engine with its double overhead camshafts has been lighting enthusiasts' fires since 1948. It was Jaguar that showed the world what disc brakes could do, at Le Mans. Jaguars had four-wheel disc brakes when Detroit's engineers were insisting that they couldn't be made to work on passenger cars. Jaguars have always represented a lot of science and sex for the money.

Certainly, the car's appearance is special. Nothing else looks like a Jaguar. My own prejudice always held that earlier XJs looked a little squished—the roof was too low—and I felt that the coarse-mesh grille texture was out of keeping with the rest of the car's elegant physiognomy. Now they've fixed it. The Series III has a noticeably higher roofline with greater glass area, which does much to lighten and brighten the car, both inside and out, and the new vertical grille bars lend a classical touch. The stamped wheel covers are unfortunate, however. If Cadillac wants to hang Pep Boys wheel covers on the Seville, that's their business, but a car as nice as this cries out for something really special. Also, the whitewall tires have to go. The ones on our Jag were Dunlop SP radials that worked well, represented a nice compromise between high performance and quiet ride, and featured a tough-looking tread pattern, but the dated whitewall did the entire car a disservice. Aside from these minor complaints, our blood-red test car looked like a million dollars. The Jaguar is one of a very small handful of modern cars with real character.

The inside story on the Series III XJ is much like the outside. Nothing revolutionary, just worthwhile improvements. When you open the door you'll notice new, recessed door handles and new, centrally controlled door locks. Once inside, the first thing to impress you will be the new seat—still leather-covered, but firmer, with a more aggressively contoured backrest and a more sophisticated range of adjustment, including a new lumbar support with an inch and a half of travel. Fore-and-aft adjustment is manual, effected by a release at the front of the seat, while backrest angle is controlled by a lever-operated friction lock, which never seemed to offer exactly the angle one was looking for. A valuable addition is a simple, electrically operated adjustment for height, especially useful when we couldn't get the backrest where we wanted it. Our tendency was to drive with the seat quite low, the backrest more vertical than we'd set it in most cars, and the telescoping steering column pushed all the way in. This put everything in reach, offered a fair compromise for best visibility, and seemed most comfortable. Our longest drive was about 500 miles—New York to New Pittsburg, Ohio—and it was a piece of cake, largely because of the very comfortable, cocoonlike interior.

Another useful addition on the Series III is cruise control, in this case one of the best we've ever used. Turn it on with the master switch on the console, then operate it with the button in the tip of the turn-signal stalk. If you're trying to stay a fixed distance behind the guy up ahead with the Fuzzbuster on his dash, you just set your speed at, say, 75, and when you see that he's edging away from you, match your speed to his with little taps on the button, then relax. The control works with micrometer precision, and shuts off with a touch of the brake pedal.

Some day, my boy, this will all be yours

• Okay, here's what we do. We find Max Hoffman, wherever he is—Palm Beach, Beverly Hills, Bavaria—and we get him to buy us Jaguar. Not *a* Jaguar, but the entire Jaguar operation, *all* of it. Some of Jaguar's palmier days in the American market were back when Max Hoffman was importer, distributor for a large part of the country, and dealer in a couple of key markets. Max may have made one of his very first millions on Jaguar. He's exactly the right man to help us in what will surely be one of the greatest automotive turnarounds since they chased Billy Durant out of General Motors. There is no earthly reason—given Max Hoffman's money and commercial acumen, and our zeal and dedication—that we can't turn Jaguar into the greatest thing since the Golden Arches. The cars are lovely. It could be argued that they'd be better if they were assembled by Oriental workers, but as a conception they are flawless. All Jaguar really needs for success is to be its own car company again, safe in the hands of the board I've just nominated. We'll be so rich that Arabs will vie with one another just to live next door to us in London.

Very well, you may say, if the Series III Jaguar is such hot stuff, why doesn't it sell like Mercedes-Benz or BMW? Well, the reason for that is something called Donald Stokes. Donald Stokes is the non-scientific term for the disease that afflicts BL, known here as JRT (Jaguar Rover Triumph), formerly known as British Leyland. If BL goes down the tube in the next year, taking such automotive wonderfulness as Jaguar, Rover, and Triumph with it, there'll be nobody to blame but Donald Stokes. He mortgaged the future of those great marques for short-term profits, just so that he might look like a no-nonsense man of action in the British daily press. And for this the queen of England made him a baron.

Rumbles from the U.K. warn that BL could go under sooner than anyone has realized, which would be a political and economic disaster for Great Britain, and even worse for the automotive-enthusiast community. This is not hearsay. The hired-gun chairman of BL, Sir Michael Edwardes (*sic*), with only a few months to go before his three-year stint is up, has remarked publicly that the ancient and honorable British institutions now represented by Jaguar, Rover, and Triumph could be sold or bankrupt by the time he's gone. (One wonders why Her Majesty's government would have entrusted this tottering enterprise to a man who doesn't know how to spell his own name, but there you are.)

There is a reverse synergism at work in BL. The whole is less than the sum of the parts. Jaguar, Rover, Triumph, even BMC (home of Austin, Morris, MG, Riley, and Wolseley) were getting along reasonably well before old Stokes folded them all into British Leyland, certainly better than any of them has done since. Everybody blames British labor for BL's troubles, but it would be silly to suggest that the Stokes legacy of management by thundering herd, never-plan-beyond-this-afternoon, and never-make-an-investment-when-you-can-cut-somewhere hasn't taken its toll. I know that there are British companies, staffed by British workers, that operate on a regular basis and turn a profit. I also know that the Series III Jaguar is a superb product with worldwide sales potential that would bring tears to Max Hoffman's septuagenarian eyes. If Mr. Hoffman will buy us Jaguar, and if other, similar groups will buy Rover, Triumph, and MG, Her Majesty's government will be off the hook, and the British automotive industry could be up and around again. And if it should turn out that the problem *is* British labor, then by God we'll build 'em in Shanghai.

—*D.E.D., Jr.*

JAGUAR XJ6

ACCELERATION standing ¼ mile, seconds

ROVER 3500
JAGUAR XJ6
CADILLAC SEVILLE
MERCEDES-BENZ 280E

13 14 15 16 17 18 19 20 21

BRAKING 70-0 mph, feet

JAGUAR XJ6
ROVER 3500
MERCEDES-BENZ 280E
CADILLAC SEVILLE

170 180 190 200 210 220 230 240 250

FUEL ECONOMY EPA estimated mpg

MERCEDES-BENZ 280E
JAGUAR XJ6
ROVER 3500
CADILLAC SEVILLE

2 4 6 8 10 12 14 16 18

CURRENT BASE PRICE dollars x 1000

ROVER 3500 (estimated)
CADILLAC SEVILLE
JAGUAR XJ6
MERCEDES-BENZ 280E

0 5 10 15 20 25 30 35 40

INTERIOR SOUND LEVEL dBA

CADILLAC SEVILLE
JAGUAR XJ6
ROVER 3500
MERCEDES-BENZ 280E

70-mph cruise
Full-throttle acceleration

60 65 70 75 80 85 90 95 100

JAGUAR XJ6

Importer: Jaguar Rover Triumph, Inc.
600 Willow Tree Road
Leonia, New Jersey 07605

Vehicle type: front-engine, rear-wheel-drive, 5-passenger, 4-door sedan

Price as tested: $25,000

Options on test car: none

ENGINE

Type: 6-in-line, water-cooled, cast-iron block and aluminum head, 7 main bearings
Bore x stroke: 3.63 x 4.17 in, 92 x 106mm
Displacement: 259 cu in, 4240cc
Compression ratio: 8.1:1
Carburetion: Lucas Bosch L-Jetronic fuel injection
Valve gear: chain-driven double overhead cams
Power (SAE net): 176 bhp @ 4750 rpm
Torque (SAE net): 219 lbs-ft @ 2500 rpm
Redline: 5000 rpm

DRIVETRAIN

Transmission: 3-speed, automatic
Final-drive ratio: 3.07:1

Gear	Ratio	Mph/1000 rpm	Max. test speed
I	2.40	9.5	48 mph (5000 rpm)
II	1.46	15.6	78 mph (5000 rpm)
III	1.00	22.8	114 mph (5000 rpm)

DIMENSIONS AND CAPACITIES

Wheelbase: 112.8 in
Track, F/R: 58.3/58.9 in
Length: 200.5 in
Width: 69.8 in
Height: 52.2 in
Ground clearance: 5.0 in
Curb weight: 4120 lbs
Weight distribution, F/R: 53.8/46.2%
Fuel capacity: 25.4 gal
Oil capacity: 9.0 qt
Water capacity: 8.8 qt

SUSPENSION

F: ind, unequal-length control arms, coil springs, anti-sway bar
R: ind; 1 fixed-length half-shaft, 1 trailing link, and 1 lateral link per side; 2 coil-shocks per side

STEERING

Type: rack-and-pinion, power-assisted
Turns lock-to-lock: 3.2
Turning circle curb-to-curb: 38.0 ft

BRAKES

F: 11.8-in dia vented disc
R: 10.4-in dia inboard disc
Power assist: vacuum

WHEELS AND TIRES

Wheel size: 6.0 x 15 in
Wheel type: stamped steel
Tire make and size: Dunlop SP Sport, ER70/VR-15
Test inflation pressures, F/R: 27/30 psi

INTERIOR SOUND LEVEL

Idle: 51 dBA
Full-throttle acceleration: 75 dBA
70-mph cruising: 71 dBA
70-mph coasting: 69 dBA

PERFORMANCE

Zero to	Seconds
30 mph	4.0
40 mph	5.9
50 mph	7.7
60 mph	10.7
70 mph	14.1
80 mph	18.2
90 mph	25.0
100 mph	32.1

Standing ¼-mile: 17.9 sec @ 78 mph
Top speed: 114 mph
Braking, 70–0 mph: 204 ft
EPA estimated fuel economy: 15 mpg

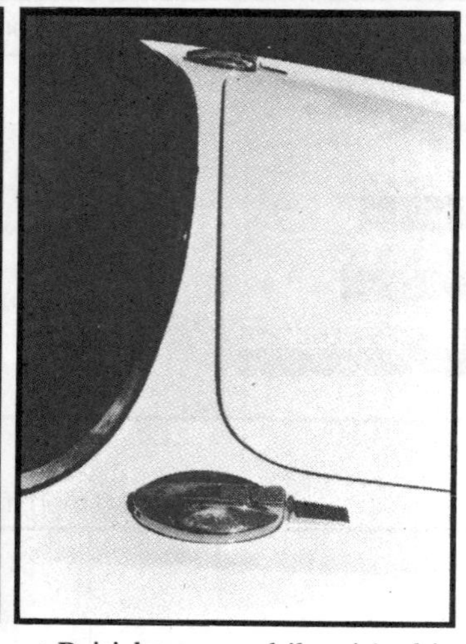

Both outside mirrors are electrically controlled now, and very logically too, by a pair of directional switches mounted at the front corner of the driver's window, exactly where a manual control would normally be—no hunting around the panel for the right-hand-mirror control. Other electrical switches are a little harder to utilize, at least until one is thoroughly accustomed to the instrument-panel layout, and the radio controls are hopeless. The radio/cassette player itself is terrific, but the markings are minute, the controls tiny, with the result that one should stop the car for anything more complicated than popping in a cassette or changing a station. The instruments are as good as the radio controls are confusing, however, being white on black, placed exactly where you'd have put them yourself, and framed neatly by a redesigned steering wheel that doesn't block your view of anything.

The XJ has two fuel tanks holding a total of 25.4 gallons, and these have to be filled separately, one filler on top of each rear fender (note: both gas caps and the deck lid lock when the central locking system is triggered). Two tanks are a drag when the temperature is near zero and you're in a self-service gas station. We also found that it can be a fright when you're sailing along at 80 and a tank runs dry. Running out of gas somewhere in the boonies is a heart-stopper, even when your mind is telling you that there's another tank available at the touch of a finger. Range with both tanks full is nearly 500 miles, figuring twenty miles per gallon. We never went that far, partly because there's always some other reason to stop, and partly because it was tough to trust the gauge twice—once for each tank.

If ever there was a car to refute the myth about British automobiles, it's this one. It always started on the first twist of the key, even in below-zero weather; the Borg-Warner automatic transmission was smooth as glass; every component did its job exactly as it was supposed to; it was a joy to drive; and the quality of everything from the map light to the twin-cam engine was first-class. It might have been better with the 5.3-liter twelve-cylinder engine, I don't know. The car is so seductive with six cylinders that twelve might have gotten me to undress right out there in the snow. One pays a $2000 premium for the twelve, but I've said before that anyone who calls himself an enthusiast should own one twelve-cylinder car before he dies. My only concern is that the twelve might not be as thoroughly scienced-out as the 32-year-old six. Besides, the XJ is so nice with six cylinders that I'm hard-pressed to imagine what real benefit one might find in six more.

—David E. Davis, Jr.

COUNTERPOINT

• Driving a Jaguar bears a lovely relation to stroking a smooth and lovely thigh. I risk an onslaught of anti-porno mail because celebrating close personal contact is the best way in which to rekindle your familiarity with the most fluid of cars. Granted, there are small harshnesses that occasionally filter through the Jaguar, but they can be likened to the somewhat less smooth textures to be found elsewhere on the human body, and they, too, are no less pleasurable than the smoothness. Suffice it to say that, for the Jaguar, too, nothing quite compares with a laying on of hands.

For this, you are rewarded with a stretching response and the faintly mechanical sighs of warmly obedient reaction. A Jaguar is as comforting as a water bed, but its uses are infinitely more controllable. Indeed, a Jaguar is the antithesis of marginal control, yet its control is accomplished with feathery finesse. A Jaguar simply assumes a receptive and yet highly participatory role in the sensual and free-thinking act of travel. Almost nothing is too great to ask of a Jaguar, and it gladly shares the very special moments of closeness. It gives as good as it gets.

—Larry Griffin

Unlike some of my colleagues, I can't quite convince myself that the XJ is the best thing since sliced bread. Not that I don't love it dearly. It's a sumptuous motoring feast, a back-roads Rolls that's sinewy, nimble, measured, and concise in its every move and response. And yet it is also uniquely silken and supple, in a way that eludes the best German and American luxury cars. The XJ is fine Burgundy, full-bodied, overflowing with character, and teeming with fascinating nuance.

The only problem with the XJ is that there's always someone around to ruin the experience for you. Everyone can recite a Jaguar horror story, a tale of woe about someone's XJ that only ran three days out of seven. And even your Aunt Millie knows the litany of hard-to-find dealers and ultra-expensive parts.

And that's too bad, because for twenty-odd grand you ought to be able to buy a sedan that's as wonderful as the Jag *and* as reliable as a Timex. For all I know the Series III version is. But I wouldn't advise any friend of mine to risk his rapidly shrinking dollars finding out for sure.

—Rich Ceppos

Pity the poor Jaguar, burdened, as it is, with not one but two legends. Empathize for a moment. How would you like to go through life known as one of the world's greatest lovers, but unable to perform much of the time because you're laid up with various infirmities?

I don't have any statistics, any hard facts to back up Jaguar's reputation for mechanical miseries bestowed upon owners, but will go with the theory that if something is repeated that often, there's some truth involved. I can attest to the validity of the Jaguar's being one of the world's great motorcars. I've just stepped out of our Series III, and it is, in a simple word, wonderful. It's a marvelous road-going device that manages to combine the traditional Jaguar fit, feel, and finish with current high-tech mechanicals. It's quiet, precise, sure-footed, responsive. It's a four-door sports car and a small limousine. That side of the legend, I think, is worth risking the other. *—Mike Knepper*

JAGUAR XJ6 SERIES III

A lovely sedan goes through some changes

PHOTOS BY DOROTHY CLENDENIN

FOR MORE THAN 50 years Jaguar has been offering a line of luxurious cars with sporting performance, cars that have been much admired and always a joy to drive. Usually understated in appearance, Jaguar's sedans have always had that undefinable quality of style and class that many manufacturers seek but few attain. The latest model to come our way for testing is the XJ6 Series III, which we reviewed briefly in June 1979, and it upholds the tradition well. However, it incorporates certain styling alterations and detail changes that are somewhat puzzling to us because, in certain cases, they seem to be changes for the sake of change rather than for more positive purposes and don't really contribute much to the appearance of the car or to the driving pleasure. Ergonomic problems with the electric mirror controls and the cruise control switches are two examples.

From the appearance standpoint, the roof line has been raised by 1.5 in. to give a greater glass area, increased head room and improved outward visibility, particularly for the rear seat passengers. The roof is less rounded and more angular when viewed from the side, which tends to give the profile a generally more fashionable appearance. The rear window is flatter, with a slightly wider roof pillar, although the parcel shelf space has not been sacrificed. The windshield pillars have been given an additional 3-in. rake forward, so the glass area is bigger. The vent wings have been removed and the one-piece front side windows have additional curvature because the roof is narrower. These various body changes are not very apparent to the casual observer and are, of course, evolutionary rather than revolutionary. We were a bit disappointed to find that the level of wind noise has increased considerably over the last XJ6 tested, being four decibels higher at a steady 70 mph.

At one time the XJ6 and XJ12 series of cars was offered in long- and short-wheelbase versions but only the long-wheelbase cars are now offered in the U.S. Unfortunately, the superb 12-cylinder engine is no longer available in the U.S.

Other external changes include a new grille with vertical bars and a center rib replacing the former grid pattern, flush door handles, bigger taillight assemblies, and an optional electrically operated sunroof. On the whole, the restyling is successful within its limits, although our particular test car was finished in white, which is generally not a good color for cars and certainly doesn't show the Jaguar off to its best advantage. Unfortunately, the beautiful Jaguar wheels have now been replaced by chrome discs.

The interior of the car has the luxurious appearance that we expect from expensive English cars and it is done in a particularly restrained manner without the touches of opulence that other manufacturers would probably apply. The interior reeks of quality with leather where there should be leather, walnut veneer on the flat surfaces and a wool headliner.

set out in a vinyl case.

The car has an abundance of instruments, which are easy to read and, apart from the general aura of luxury, there are a number of nice functional touches, such as the deep, classic footwells, the glovebox lid that folds flat with a fold-up mirror and good console storage. Also there are map and storage pockets in all four doors and in the seatbacks. The automatic climate control works well, except that the compressor works all the time, and can be assisted by high and low fan settings, and there are outside air vents at each footwell.

By today's standards, the 6-cylinder twincam 4.2-liter engine is big and it will go beyond its 5000-rpm redline in the drive range, which is in excess of 120 mph. Jaguar has always paid consider-

Electric controls for the outside mirrors are conveniently located but, unfortunately, did not always work properly during our test.

The driver's seat created considerable disagreement among the various staff members who drove the car. It has the normal forward and back movement and a lever control for the backrest. In addition, it has an electric control on the corner of the seat which raises and lowers the whole seat in a vertical plane. Unless you know the electric control is there, it is difficult to get comfortable. There is also a 3-position control on the side of the backrest for lumbar support, but it is too severe and no one would want it in the fully supporting position. But the basic problem is that the seat padding has been changed, resulting in too much lower back support. Why Jaguar should elect to change the driver's seat is a puzzle, because no one we know ever found a Jaguar sedan uncomfortable to drive and the new seat adds nothing to the comfort. The rear seat is excellent with head room and outward vision improved by the increased height of the roof, and plenty of leg room.

The trunk is large but rather flat, although one can stand grocery shopping bags in it. It contains a neat little set of tools all

AT A GLANCE

	Jaguar XJ6 Series III	BMW 528i	Mercedes-Benz 280E
List price	$25,000	$20,150	$26,466[1]
Curb weight, lb	4070	3400	3565
Engine	inline 6	inline 6	inline 6
Transmission	3-sp A	4-sp M	3-sp A
0–60 mph, sec	10.6	8.2	11.4
Standing ¼ mi, sec	18.2	16.7	18.6
Speed at end of ¼ mi, mph	78.5	86.0	76.5
Stopping distance from 60 mph, ft	162	158	155
Interior noise at 50 mph, dBA	67	66	68
Lateral acceleration, g	0.704	0.737	0.687
Slalom speed, mph	51.9	57.2	56.5
Fuel economy, mpg	13.5[2]	22.0	17.0

[1]Price on west coast; $26,193 on east coast
[2]See text

able attention to the appearance of its engines and the engine compartment is exceptionally clean and neatly laid out. The engine starts easily from cold and shows no tendency toward driveability problems. It is extremely smooth, but not completely free of mechanical sounds, and its torque curve is well matched to the automatic transmission. However, presumably for emission-control purposes, the transmission shifts unless you manually keep the lever in 1st gear, and a heavy throttle foot is required to shift down into low. In addition, the shift lever is somewhat balky and vague so the car is best driven by selecting the drive range and leaving it there.

Some people may regret the passing of the V-12 in the U.S. market because of its silky smoothness and exotic design. However, it carries a heavy fuel consumption penalty and Jaguar's 6 is one of the smoothest so there is no real necessity for the 12. Unfortunately, our test car was suffering from a minor fault in the fuel-injection system, so we recorded only 13.5 mpg. However, we feel that a realistic figure would be about 15–16 mpg.

On the road the car is delightful, the steering is positive but the effort is a little on the light side for some of our staff members, and the suspension is an excellent compromise, being sufficiently soft to soak up road irregularities but still firm enough so the handling qualities are not sacrificed. It's a most exhilarating car to drive fast, with balanced understeer its major handling characteristic and just mild complaint from the front tires to let you know when you are pressing it hard. We rated the brakes overall as good, with minimum stopping distances of 162 ft from 60 mph and 301 ft from 80 mph.

Summing up the Jaguar XJ6 Series III is very difficult if one is to be completely fair to both the potential buyer and the manufacturer. British Leyland has a poor reputation in the U.S. for quality control generally, and we must admit that we experienced various problems with the four different Jaguars that passed through our hands during this test. One thing we always try to avoid saying about our test cars is, "On the one hand this, and on the other hand that," but in the case of the Jaguar it is inevitable because, on the one hand, it has some outstanding qualities that are equal or superior to some of the finest but, on the other hand, it is haunted by a reputation for poor quality control and indifferent service and parts facilities, which we have to believe is warranted. Obviously, it is much easier and more pleasurable for us to praise a car wholeheartedly, but unfortunately in the case of the Jaguar our praises have to be tempered with our reservations.

PRICE

List price, all POE	$25,000
Price as tested	$25,000

GENERAL

Curb weight, lb/kg	4070	1848
Test weight	4130	1875
Weight dist (with driver), f/r, %		53/47
Wheelbase, in./mm	112.8	2865
Track, front/rear	58.3/58.9	1480/1495
Length	199.5	5067
Width	69.7	1770
Height	54.2	1377
Trunk space, cu ft/liters	14.3	405
Fuel capacity, U.S. gal./liters	25.4	96

ENGINE

Type		dohc inline 6
Bore x stroke, in./mm	3.63 x 4.17	92.1 x 106.0
Displacement, cu in./cc	258	4235
Compression ratio		7.8:1
Bhp @ rpm, SAE net/kW		176/131 @ 4750
Torque @ rpm, lb-ft/Nm		219/297 @ 2500
Fuel injection		Lucas Electronic
Fuel requirement		unleaded, 91-oct

DRIVETRAIN

Transmission	automatic; torque converter with 3-sp planetary gearbox
Gear ratios: 3rd (1.00)	3.07:1
2nd (1.46)	4.48:1
1st (2.40)	7.37:1
1st (2.40 x 2.0)	14.74:1
Final drive ratio	3.07:1

CHASSIS & BODY

Layout	front engine/rear drive
Body/frame	unit steel
Brake system	11.2-in. (284-mm) vented discs front, 10.4-in. (264-mm) vented discs rear; vacuum asst
Wheels	steel disc, 15 x 6JK
Tires	Dunlop SP Sport, ER/70VR-15
Steering type	rack & pinion, power assisted
Turns, lock-to-lock	3.3
Suspension, front/rear	unequal-length A-arms, coil springs, tube shocks, anti-roll bar/lower A-arms, halfshafts, trailing arms, dual coil springs, dual tube shocks

CALCULATED DATA

Lb/bhp (test weight)	24.3
Mph/1000 rpm (3rd gear)	2650
Engine revs/mi (60 mph)	22.6
R&T steering index	1.27
Brake swept area, sq in./ton	207

ROAD TEST RESULTS

ACCELERATION

Time to distance, sec:	
0–100 ft	3.7
0–500 ft	10.0
0–1320 ft (¼ mi)	18.2
Speed at end of ¼ mi, mph	78.5
Time to speed, sec:	
0–30 mph	4.0
0–50 mph	7.9
0–60 mph	10.6
0–70 mph	14.3
0–80 mph	19.0
0–90 mph	26.0

SPEEDS IN GEARS

3rd gear (5000 rpm)	117
2nd (5000)	80
1st (5000)	49

FUEL ECONOMY

Normal driving, mpg	13.5 (see text)

BRAKES

Minimum stopping distances, ft:	
From 60 mph	162
From 80 mph	301
Control in panic stop	good
Pedal effort for 0.5g stop, lb	25
Fade: percent increase in pedal effort to maintain 0.5g deceleration in 6 stops from 60 mph	20
Overall brake rating	good

HANDLING

Lateral accel, 100-ft radius, g	0.704
Speed thru 700-ft slalom, mph	51.9

INTERIOR NOISE

Constant 30 mph, dBA	63
50 mph	67
70 mph	74

SPEEDOMETER ERROR

30 mph indicated is actually	30.0
60 mph	60.0

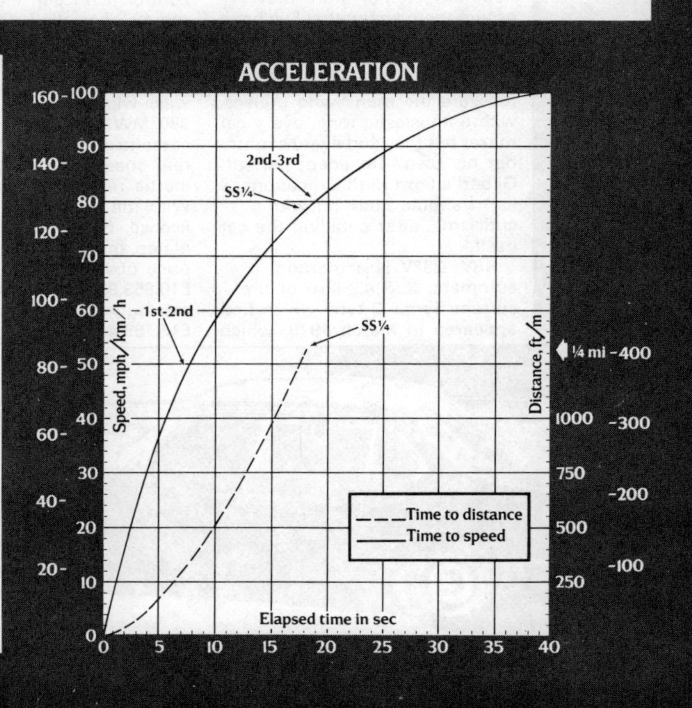

LONG-TERM REPORT

Jaguar XJ6 Automatic
(12,000 miles)

By Michael Scarlett

A superb and barely rivalled motor car seen in one of the many viewpoints from which it looks so pleasing – here particularly feline

Coventry-made marvel

In a car like this, the faults hardly matter

I HAVE HAD to pinch myself on several occasions. No car made is perfect, of which anyone, particularly a bloke involved in *Autocar* Road Tests for 13 years, is only too well aware; and this one has faults and failings. But, in spite of those faults, and in spite of a well-tempered case-hardening of fine-grained cynicism acquired through years of saddening experience, to be given the charge of a Jaguar XJ6, to have the chance of living with the car, treating it as one's "own" – it borders on the edge of privilege.

At which point already nauseated readers who have got so far may want to turn to hopefully sterner stuff. May I assure them again, that I've been driving other people's motor cars in order to test them for too long to become infatuated. Perhaps, to reassure the man "who praises, with enthusiastic tone, every car maker but this, and every country but his own" (to adapt part of Gilbert's Lord High Executioner's list) I should start off with criticisms, after detailing the car itself.

Below: The much better quality electric aerial now fitted to XJ models. Bottom: Distinctive Series III roofline is obvious in this side shot, with the slightly thicker rear quarter panels

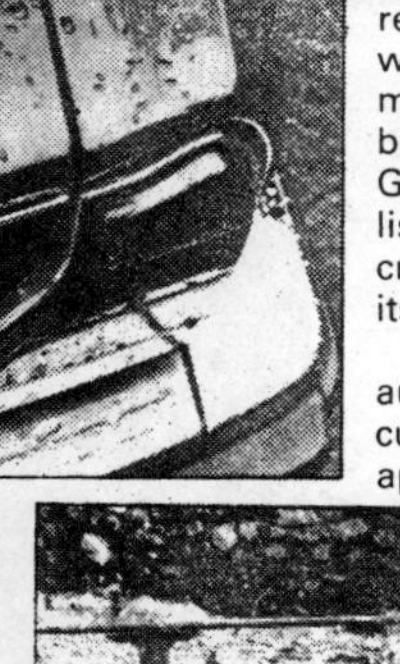

HRW 287V is a maroon automatic XJ6 4.2-litre of the current Series 3 type which first appeared in March 1979, which means that it has the electronically controlled L-Jetronic Bosch fuel injection 208 bhp engine and Borg Warner type 66 three-speed automatic transmission set in the Pininfarina-modified body with the more shallow-raked windscreen, longer roof, steeper back light and more pronounced tumblehome. It began its life with *Autocar* as our Road Test car, arriving on 2 November 1979 with 2,366 miles on its distance recorder, and most of the possible extras available for the XJ6. At then-current prices, it sported the excellent Delanair automatic air conditioning system (£1,109.43), AE cruise control (£287.86), electrically remote-controlled door mirrors (£103.63), electrically motored seat cushion rake adjustment (£168.51), dipped lamp headlamp wash-wipe (£184.23), a Philips 880 MW/LW/stereo FM cum cassette player with front and rear speakers (£210.20) and inertia reel *rear* belts (£66.45). With the addition of £115 for licence, delivery and number plates, this lot put the catalogue price of £14.608.56 up to £16.853.95. Since then, the catalogue price has gone up to £15.798.41 (8.1 per cent more) and the total (including that £115) to £18,043.72. It seems an awful lot of money when one remembers the first automatic XJ6 of September 1968, which you could buy in standard un-extra'd form for £2,397.

Inflation is responsible for much of the difference, of course; looking at nearest equivalent Mini prices, the Mini City costs 4.41 times the price of the 1968 Mini 850 Super de luxe 850, and of course one gets rather more in a Series III Jaguar 4.2, even without the extras, than you did in the original, so the fact that it costs basically at least 6.6 times more is tolerably forgiveable. But one is still tempted to ask what has happened to that Jaguar reputation for competitive pricing – until you look at the competition from abroad, remembering that it is nowadays unhindered by tariff barriers thanks to the EEC. The nearest equivalent capacity and class cars to the Jaguar's £15,798 today are the BMW 735i (3,453 c.c.) at £16,535, and two Mercedes, the new 380SE at £18,400 and 500SE at £20,950. All three are excellent cars in their class, but – and this is a statement devoid of prejudice, derived purely from objective judgement, which (prejudice briefly appearing) I would like politely to ram down the throats of all the British idiots who utter or believe foolish fashionable generalisations about foreign cars being inherently better than British ones – neither the BMW or each of the two Mercedes are overall the equal of the Jaguar.

Leaving my soapbox again, HRW 287V ended its Road Test duties at 5,331 miles on 8 December, so that this traditionally 12,000 mile long term test is deemed to beg from that point, up to arou 17,499 miles, allowing for t average mileometer exaggeration of 1.4 per cent. can argue that, regrettably, denies us any experience o delivery and running-in per faults, but it does mean tha car's life began with most untypically hard driving, particularly during the takir test performance figures and investigation of ultimate roadholding and handling behaviour.

So, as promised, what ha gone wrong in that mileage Mostly little things, which underline the ominous-for-s recent remarks of Jaguar's chairman, John Egan, about some component or mater suppliers. The electrical ae would not retract more than way by the time of the 6,000 service (more of this anon) industry has been suffering some time of late from poo quality drive vee-belts, wh suspect was the real reason w had to tighten the air conditioning compressor be twice between the 6,000 mi service and the 12,000 one, a it and the two other belts (for alternator, and fan cum wat and steering pumps) were replaced. Incidentally, belt tightening as a first aid is m exceptionally easy by the w Jaguar provide a screwed s instead of the usual plain so arm, which does away with need for tyre levers to hold new adjustment as you tigh up. (There is a similar piece typical Jaguar thoughtfulness both the other belt adjustmen

At 7,100 miles, grease wa spotted on the nearside fo wheel hub, suggesting a leal

əal, but this did not continue, so ıaybe it had simply been verfilled. One item we no longer st in *Autocar* Road Tests is the umber of greasing points, ecause grease nipples have rgely disappeared from most ırs as more makers adopt so-alled sealed-for-life bearings of l sorts. Maybe I'm a stick-in-the-ly-mud cynic, but no one is ever ıally clear about how long that 'e is – which is why, if I truly wned HRW 287V, I'd be happier bout the existence of as many as ıe 17 grease nipple points on the J.

Bearings for suspension, eering, rack adjuster pad and rop shaft universal joints all ave grease nipples, which if tended to as recommended ıust lengthen life, and make it ıore likely that any instance of ndue wear or looseness is potted in time. That must be less kely with joints that do not quire or permit any greasing. ne cannot however help ondering if 6,000-mile intervals 'e a bit too frequent.

With colleagues Warren Allport nd artist Vic Berris, I used the aguar to make the first of two sits to Rolls-Royce at Crewe to arn about the new SZ Silver pirit range. We gave R-R's ondon-based public relations ıanager Dennis Miller-Williams lift, and it was therefore mildly nbarrassing to discover on that ip a small but discernible click om the rear on any marked ıange of throttle opening. aving, as any owner might, got sed to the car and into the habit f driving it comparatively moothly – not at all difficult in ny case with such typically-for-aguar superb throttle control nkage – I hadn't provoked the oise. Warren had not at that age driven it before, and it was hile he was getting used to it ıat the sound was aroused; you ad to be moderately abrupt with ıe right foot to provoke it. I still ave not established for certain ıe exact cause, but as a small ut just audible amount of final rive whine on drive, not over-ın, had appeared by 15,000 ıiles, I obviously suspect some lay in the rearward part of the ansmission.

That great boon, central ocking, caused some uneasiness etween 13,000 and 15,000 miles. worked perfectly satisfactorily s usual when worked from the river's door, but could ıisbehave oddly if one unlocked ıe car from the left hand front oor, say first thing in the ıorning, in order to put omething inside from the avement. When one then went ɔ the driver's side, as soon as ne pulled the outside release to pen the door, the central locking romptly locked the entire car, nly to behave properly when the ey was used on the same side. he fault occurred irregularly ntil it was dealt with at 15,000 ıiles.

In the air conditioning system, one normally hears occasional small whirring noises as the automatic control dictates a temperature correction. These sounds are not objectionable – but at an indeterminate point before 15,000 miles, I became aware in low speed town running of a tendency for the system to make a continuous but varying tone musical note – not loud, but noticed because the car is then otherwise so outstandingly quiet.

For some reason – I suspect unintended misuse by someone – the walnut veneered lid of the glove locker became stiff enough to open for strangers to believe that it was locked; resetting at 15,000 miles sorted that one out.

One of the advantages of having the headlamp wash-wipe system is that the wash reservoir, which is common to both headlamp and screenwash, seems for all practical purposes inexhaustible. It is housed in the left hand wing forward of the wheel arch, and holds 1.63 gallons. The only irritation here is that its cap, a thin yellow plastic item, tends to crack so that it won't stay on, although the reservoir's shape and size seems to prevent any undue loss of fluid.

Paint problems

Any car that is often driven fast – and that certainly applies to this one – will suffer from stone-chipped paint. That has obviously been so for the Jaguar, but the appearance of some surface paint chips in places not so easily struck by road stones has made me wonder how resilient the paint is, particularly when the grapevine suggests that around the time of HRW 287V's birth, there were some quality control problems with paint supply.

So much for failures or failings. There are other variously important disappointments of course. The XJ6 has always been an outstandingly quiet car in all respects, bettered only by its V12 big brother, which wins simply because that most superb of all big engines never gets noisy within its remarkable rev range, whereas the six-cylinder XK engine still begins to become a little fussed near the 5,000 rpm red line. My particular car makes me wonder whether the Series III XJ is quite as good a suppressor of wind noise as the Series II body; there is a little aerodynamic noise from around the front doors, which of course suggests the usual reason for such sound, which is rarely the car's shape as much as plain door sealing. I should emphasize that the level is one that in most other cars one would call low.

The car's marvellous handling and roadholding, plus its tyre life, leaves me in a small quandry about its tyres. They are Pirelli P5 in the standard size, 205VR/70-15in. Their virtues are undeniable and highly endearing to any keen driver, especially the one who is in a hurry. They give unusually good response, better than the norm for the XJ, and this is coupled with a better than usual reluctance to squeal, which has the practical advantage of allowing one to exploit the car's superbly unfussy flat cornering abilities much further without irritating others with tyre noises. They have excellent adhesion in both dry and wet, and don't aquaplane readily whilst less than half worn. In spite of good grip, they are remarkably long-lived; as I explained at the start, this car has been through the full Road Test routine, which includes a great deal of very severe cornering which is very hard on tyre life. And its outstanding cornering abilities encourage every driver to belt it through bends wherever reasonable. In spite of that, the likely life of the tyres projected from their wear at 13,600 miles is at least 24,000 miles front, and 21,000 miles rear. Because of the reducing effect of the much higher Road Test wear rate as the overall mileage rises combined with the fact that a tyre's wear tends to decrease later in its life, the projected life increased towards the end of the 12,000-mile test period to 27,000 miles front, and slightly *more* for the rear.

The quandry for me is that the one price one pays for the P5 is a definite increase in road noise compared with the Dunlop tyre originally designed for the XJ. Road noise levels for the car on P5s can be fairly described as only better than average instead of outstandingly quiet. It is, however, only fair to point out that this is a matter of opinion; my colleague John Miles was able to conduct a back-to-back test of the P5 against the Dunlop, returning saying that whilst there is a noise advantage to the SP Sport, it wasn't enough to offset the improved response of the Pirelli.

Continuing on the handling tack, one gets used to the frequently criticised lightness of the Jaguar's steering, continuing to wish only that it was around one third higher-geared. Contrary to what some people say, it does have feel; there is actually a nice degree of kickback (this I suspect in fact improves as the car's mileage rises), and once you learn to drive the car with the fingertips, even *in extremis,* it gives good messages of what the front wheels are doing.

I have few complaints about the performance, which has actually improved slightly during the test period. All right, so I would like it to accelerate a little better – it is remarkable how used one can become to a very refined automatic which gets to 60 mph in 10sec, and to 100 mph in just under 30sec – but Jaguar will only achieve that with a brand new car which weighs a lot less than the present one's 34.6cwt. The fuel injection gives effortless unfussy starting and driveaway in all weathers; on a cold day, when there is ice over the entire car, I revel in the way one can open the door, put the key in the ignition, and start so quietly and confidently, with none of the usual high revs of an automatic choke, without touching the accelerator pedal. Our car will now achieve a mean maximum speed of 130 mph, against the Road Test 127. More

Modification on later Series III cars is this large thumbwheel under the radio, controlling how much lower the face level air temperature is; automatic air conditioning system is a great blessing in all weathers

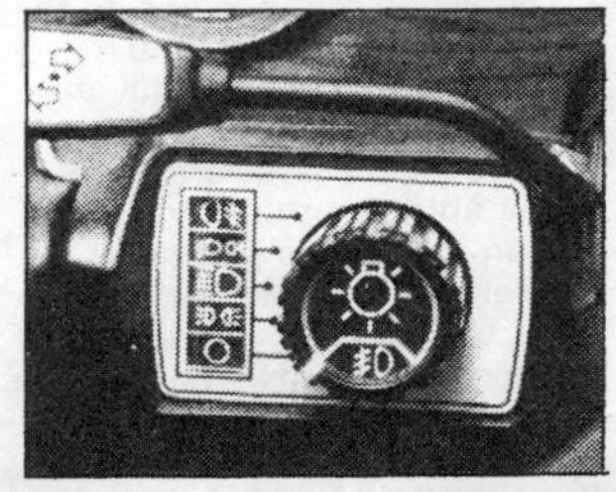

Left: All-in-one combination light switch drives anyone less than very familiar with the car mad; but owners who become used to it find it satisfactory

Facia layout is a delight to the eye, and works well

practically, its abilities at cruising at a relaxed 100 mph are so subtly competent that it is a sort of strain holding the car down to the 80 mph motorway speed to which the absurdity of Britain's crazy 70 mph limit in effect equates.

It responds so smoothly to the demands of the driver, the only exception being the slightly unreliable standard of gearchange up and down from the Model 66 transmission when pressed. Sometimes both kicks down and changes up are pretty smooth; other times, there is a definite jerk, which is not up to the standards to be expected from this sort and weight of car. My other complaint about the transmission is the old but still so vitally important one we have regularly to charge against Jaguar – that dangerous selector gating arrangement, with a stop where it isn't needed, between 2 and D, and none to prevent the driver negotiating the lever past that stop on changing up from second overriding D straight into Neutral, or, almost equally easily because again of that 2-to-D detent-avoiding side pressure, into Reverse. As a simple short term improvement, all that is needed is to change the selector gating cam to provide detents between D and N, and between 1 and 2 (although that is less important thanks to the presence of an inhibitor in the gearbox), and of course to delete the detent between 2 and D. In the long term, it would be best of all if Browns Lane quietly took a leaf out of Unterturkheim's book, and provided as sweetly moving a slot-type gate, correctly arranged, as on all Mercedes automatics.

The action and position of the throttle pedal are beyond criticism, but the fact that the brake pedal is appreciably forward of the accelerator is regrettable, for the obvious reason that moving from accelerator to brake requires an extra unnecessary movement of the foot rearwards. As with the several other cars with this failing, it always surprises me how easily one gets used to it; the extra movement admittedly becomes unconscious, provided that one drives conventionally, using the right foot only. For town driving, where there is frequent slowing and stopping, I tend to use the left foot for braking – and then the brake pedal's rearward position is a nuisance which is always noticed.

That famous engine seen from the fuel injection side. Early on in our acquaintance I became mildly alarmed at a small clicking noise heard from a running engine outside the car; it's harmless, coming from the injector solenoids

Modifications

At approaching 15,000 miles, Jaguar told us that there were some small modifications being introduced on the production line which it would be worthwhile fitting to HRW 287V. These were the substitution of a much better electric aerial for the previous one, and an extra thumbwheel control for the air conditioning system. The latter essentially controls whether or not the upper level (face-level) air is made cooler than lower level air, and overall the system now provides a more noticeable gentle flow of cool air to the face, meeting the one criticism previously made of the system. Drivers more familiar with the hearty blow from well through-flow-ventilated humbler cars like Fords and Austin-Morris products at first tend to ask why the cool flow isn't stronger – but experience with the great and delightful luxury of air conditioning suggests to me that when the inside temperature of a car is automatically kept down to a reasonable and very comfortable level when it's hot outside, you don't need a strong blast; a gentle blow is just right. I can't speak for American car systems; since no chance has occurred lately to up-date myself on transatlantic standards of refined air conditioning, but the Jaguar one has no superior in Europe. It is quickly effective, has good control, and is usually very quiet; I say "usually", remembering that low moan mentioned earlier.

The new aerial is similar to the type used in Rolls-Royces, has proved to be a great improvement, and totally reliable. It is much faster than most at putting itself up, erecting itself in less than 2sec, and is much less sensitive than the previous one seemed to be to how often it needs cleaning or oiling.

The car has been used as a tyre test demonstrator; at 13,600 miles, John Miles who, thanks partly to his racing experience and partly to his inate abilities as a development engineer, has become *Autocar's* tyre boffin, took over HRW to try a set of Pirelli 235/60 VR15in. P6s on wider rims, which necessitated 3/16in. spacers at one end. Such very low profiles can be expected to bring losses in ride and bump-thump suppression, but I was agreeably surprised at how small these losses were – it said much for how well the suspension itself, regardless of the vital contribution of a tyre, works. There was also of course a distinct improvement in response which was marvellous, without apparently any increased tendency towards aquaplaning, so often the fatal weakness of putting wider tyres on a car.

Otherwise the car's life has been comparatively uneventful in this long term test period. I have detailed the problems thoroughly. None of them has put the car out of action, because none has been truly serious, with the not fully investigated transmission click as an arguable exception – if I owned the car, I would have insisted on a total cure – the fact that I have not is as much due to the present smallness of the noise as to anything else.

To be fair to something like an XJ6, when it is at least five times more expensive a car than you could afford yourself, you must put yourself in the fortunate position of someone who can afford it, but who nevertheless has highly critical standards. Such a machine cannot be cheap to run. The 12,000-mile service included one out-of-the-ordinary change of all three drive bel (themselves costing £11.10 including VAT), and new pads the back brakes (remember, t car had undergone *Autocar's* deliberately arduous fade test which it passed much better th any of its Continental rivals c but nevertheless the total bil came to £162.24.

Above: Replacement washer reservoir cap is beginning to crack like its predecessor. Below: Oil filler cap should be snap action like Rolls-Royce to save burning fingers

For its size and weight, the Jaguar is not thirsty. Asking a engine to accelerate over 1¾ tc of 70in. wide saloon from a standstill to 100 mph in und half a minute, and to do so through the inefficiencies of typical automatic gearbox, ar that engine is going to dema some feeding. The car returr 16.8 mpg overall in our Road T and has naturally done better more typical running since, averaging between that figur and 18 mpg according to conditions and driving. It is possible to extract as much as mpg if you are very restrained a long run, but I can't say that t is typical XJ6 driving. Oil consumption, on the face of has nearly doubled since the Road Test period, when keep the level topped up to the dipst maximum – and observing Jaguar's rightly demanded o minute delay between engin stop and taking the reading allow for oil draining) consum a pint every 800 miles. I hav since found, originally throu simply not topping up when level was halfway, that the ca oil consumption drops if you maintain that halfway mark, instead of maximum one. Consumption now averages 1,500 miles per pint, which acceptable in my opinion.

Petrol and oil consumption servicing costs – these are bores of motoring. I have kept joy – and it is nearly always unadulterated joy – of drivir and running such a car to the la It always gives me pleasure j to look at, even if I still have quite accepted the attentions the greatest of all Italian styli to the shape; the slight flatteni and angularities imposed on back of the superstructure dc improve on the original sleet curves attributed to the inspi Sir William Lyons or marry entirely happily with what is of them now. But it is still stunning to gaze upon, from almost all angles – the sort of which it is a pleasure to photograph.

Few equals

Its performance is so smoo and refined that on getting in any other car for testing purposes, I have to take a men grip and remind myself that standards must in most cases lowered, because I've been driving the Jaguar. The sweetness and perfect driveability and driveaway fro cold are nearly equalled by a fe other cars, with six-cylinder f

egroom in rear is generous, even for adults; centre armrest, here folded back, is ıst the right height for most people's comfort

ıjection engines, but not xcelled. The easy, always bedient rustle of that superb ngine is wedded to an ideal lend of steering, ride and andling, under a hardly isturbed umbrella of efinement. A passenger, bserving the unconsciously igh cornering rates at which one omes to drive the XJ6, said how rounded bends and swerves ike a big Mini". In its low roll nd excellent response to ngertip steering, plus the very ood all-round view, its abilities threading traffic neatly and fficiently are marvellous – there always the chance to take dvantage of a quickly opening pportunity without annoying ther people.

iadgets galore

The air conditioning system ıakes life in the XJ so very much ıore comfortable at all times and all weathers. I use and ppreciate the majority of its adgets, though I would be erfectly happy to have a rincess-style hand-worked seat ushion *height* adjuster instead f the very expensive electric ake-only adjuster, and could urvive without the undoubtedly xcellent cruise control, which is nly of any use on motorways ss crowded than our own.

There are a lot of nice toys tted to the car. You can half-ower the aerial on entering a ow-headroom car park, so that ery good Philips radio continues o work. Passengers appreciate ne 10sec delay switch-off for the ourtesy lamps, unless they are eeping one-year-old daughters vho you desperately don't want o wake at the end of an evening rive; there is no way of stopping ne courtesy lamps coming on in uch circumstances. Older emales are grateful for the raditional Jaguar position of anity mirror, normally folded at on the inside of the glove ocker lid. The remote control oor mirrors work well, and don't lur with vibration at any time. entral locking makes parking ne car and leaving it that much asier. The electric windows are a oon too. The horn switch is in ne right place, on the steering vheel cross bar (though some riticise it for needing a quite hard operating squeeze). I deplore Jaguar's joining the drive-on-the-Napoleonic right ISO brigade's mistaken control standard which imposes the signalling stalk on the wrong (left) side for British drivers, but love the frictionless, notchless action of both this and the wipe/wash stalk. The leather seats with, in front, their adjustable lumbar support are very comfortable, and also hold one quite well sideways.

The XJ6 is a rewarding car for the enthusiastic owner who prefers to do routine servicing, at any rate the whole of the 5,000-mile one and assuming no unscheduled awkwardnesses, the 10,000 one too. A truly excellent maintenance handbook accompanies the hardback normal handbook, and most refreshingly for today encourages what it calls a "person with average mechanical ability using normal hand tools to carry out routine maintenance himself." It claims that the "only hand tools required for owner maintenance in addition to" the excellent set "provided with the car, are a trigger oil can and a grease gun" (the automatically gripping lever sort). There is no difficulty about buying the "repair operation manual" (workshop manual), which costs a tolerable £7. The demands of my job and of a house needing an overhaul have meant that I've had to break my normal rule of servicing the car myself, but judging from odd jobs which I have done, it isn't nearly such an alarmingly complicated car to work on as first appearances suggest. This is partly due to nice little Jaguar details, like the belt-tensioning stays mentioned earlier. I would recommend any owner near enough who wants the servicing done properly by Jaguar specialists who know what they are doing to go to R. A. Creamer and Son Ltd., in Drayson Mews, Holland Street, Kensington.

The car happily is still with us, and will be reported on at least one more time later in its history. I can only say that I'd be delighted to run it for a lifetime. It represents something very special in British cars, and something paramount internationally. □

WHAT THE JAGUAR XJ6 HAS COST

Total for 12,000 miles £7,805.79. **Cost per mile** 65p
Mileage now: 17,500 Age now: 8 months
Period covered by this report: past 12,000 miles and 7 months
**Based on use over 12 months and inc. standing charges*

CONSUMABLE ITEMS	Life in miles	Unit cost	Cost per 12,000 miles
Fuel: 4-star (gallon)	17.4	£1.30	£896.55
Oil: topping-up between changes (litre)	2,640	£1.85	£8.41
Brakes:			
Front pads (Unit cost=set of 4)	24,000	£32.20	£16.10
Rear pads (Unit cost=set of 4)	15,000	£16.10	£12.88
Tyres: Pirelli P5 205/70 VR 15in.			
Front pair	27,000		£107.20
Rear pair	29,000		£99.81
(Unit cost=single tyre)		£120.60	

SERVICE and REPAIRS		
Recommended charges for service at £10.00 per hour (labour only)	£83.87 at 6,000 miles £136.00 at 12,000 miles	
Service costs incurred with our car in past 12,000 miles (inc. oil and materials)		£329.98
Repair costs incurred		–
Total maintenance costs incurred		£329.98

STANDING CHARGES	
Insurance (see note) for 12 months	£174.86 (group 7)
Tax for 12 months	£60.00
Depreciation estimate	
Cost of our car new	£14,609
Value today (car 8 months old)	£10,500
Estimated depreciation (12 months)	£6,100

SUMMARY	
Total running costs (consumable items plus Service and Repairs)	£1,470.93
Standing Charges (12 months)	£6,334.86

Note: To put all our cars on equal footing for insurance cost, the figure given above is a typical quotation for a "good risk" driver – with clean record, and car garaged in Oxfordshire, a "middle range" risk area. Full n.c.b. discount has been deducted, as has the saving for £25 excess. The actual figure given is the middle one of five quotations. Source: Quotel Motor Insurance Service.

MAXIMUM SPEEDS

Gear	LT mph	LT rpm	RT mph	RT rpm
Top (mean)	**130**	**5,250**	127	5,150
(best)	**131**	**5,300**	130	5,250
2nd	**85**	**5,000**	85	5,000
1st	**51**	**5,000**	51	5,000

ACCELERATION

FROM REST

True mph	Speedo mph LT	Time secs LT	Time secs RT
30	**31**	**3.8**	3.8
40	**41**	**5.7**	5.6
50	**51**	**7.5**	7.4
60	**61**	**10.1**	10.0
70	**72**	**13.3**	13.3
80	**82**	**17.1**	16.9
90	**92**	**22.1**	21.9
100	**102**	**29.7**	29.4
110	**112**	**38.9**	38.5
120	**122**	–	–

FUEL CONSUMPTION

Overall mpg:
LT **17.4** (16.2 litres/100km)
RT 16.8 (16.9 litres/100km)

Note: "RT" denotes performance figures for the same car tested in Autocar of 29 December 1979

Standing ¼-mile:
LT 17.4 sec 85 mph RT 17.4 sec 85 mph
Standing km:
LT 31.8 sec 102 mph RT 31.6 sec 103 mph

IN EACH GEAR

mph	Top LT	Top RT	2 LT	2 RT	1 LT	1 RT
10-30	–	–	–	–	**2.9**	2.9
20-40	–	–	–	–	**3.4**	3.3
30-50	–	–	**5.3**	5.3	**3.3**	3.3
40-60	–	–	**5.6**	5.5	–	–
50-70	–	–	**6.0**	5.9	–	–
60-80	–	–	**6.8**	6.7	–	–
70-90	**10.8**	10.7	–	–	–	–
80-100	**12.7**	12.5	–	–	–	–
90-110	**16.4**	16.2	–	–	–	–

SPECIFICATION

ENGINE	Front; rear drive
Cylinders	6, in line
Main bearings	7
Cooling	Water
Fan	Viscous and electric
Bore, mm (in.)	92.07 (3.62)
Stroke, mm (in.)	106.00 (4.17)
Capacity, cc (in³)	4,235 (258)
Valve gear	Dohc
Compression ratio	8.7-to-1
Fuel injection	Lucas/Bosch L-Jetronic
Max power	205 bhp (DIN) at 5,000 rpm
Max torque	236 lb. ft. at 3,700 rpm

TRANSMISSION

Gear	Ratio	mph/1,000rpm
Top	1.0-2.0	24.7
2nd	1.45-2.90	17.1
1st	2.39-3.78	10.3

Final drive gear	Hypoid bevel
Ratio	3.07

SUSPENSION	
Front	
–location	Independent, wishbone
springs	Coil
dampers	Telescopic
anti-roll bar	Standard
Rear	
–location	Independent, lower wishbones, fixed length drive shafts
springs	Twin coil
dampers	Twin telescopic
anti-roll bar	Standard
STEERING	
Type	Rack and pinion
Power assist	Standard
Wheel diameter	15¾ in.
BRAKES	
Front	11.2in. dia. disc
Rear	10.4in. dia. disc
Servo	Vacuum

PRODUCED BY:
Jaguar Cars,
Browns Lane,
Allesley,
Coventry,
Warwickshire (CV5 9DR)

Maroon masterpiece

Farewell to our long term test Jaguar XJ6 – still good for 130 mph at 30,000 miles

By Michael Scarlett

LIFE does go on of course but it isn't quite what it was before 5 December 1980.

For on that day, 363 after it was borrowed from them, our Jaguar XJ6 4.2 automatic long term test car was returned to its owners and makers at Browns Lane, Coventry. It had 30,032 hard-worked miles behind it, and 722 miles of the dirt of around 12 hours of near-continuous driving through much of the previous night greying its polished maroon paint. A disgraceful state in which to return such an aristocrat, but hopefully excusable, given a last minute minor marathon to knock the mileage recorder over 30,000 miles.

The car was due to go back on Friday (the 5th). During Thursday, when I had taken my wife and daughter across to stay with my parents-in-law in Suffolk, each time I had looked at the mileometer, there it was – so idiotically irritating. 29, 120 miles indicated at our Richmond home: 29,215 at Grundisburgh.

Returning to Richmond on my own on that Thursday evening, the thing said 29,310 at two minutes to 11. The run the next morning to Jaguar via the Motor Industry Research Assocation's proving ground (MIRA) near Nuneaton would add at most only 150, leaving at least 540 miles short. There would of course be the usual excuses for not getting to 30,000; pressure of work at the office, the demands of house restoration, and so on. Feeble stuff – a scurvy way to treat such a jewel of a car, even oddly ungrateful after the privilege of running such a machine for nearly a year.

Mileage boost

Since the quickest way to put on that sort of mileage is to use motorways, I set out for the beginning of M4 at Chiswick. It's a dry night, with little wind. The forecast is of mild weather, getting warmer. Quite fortuitously, the car is well enough equipped for the trip; I had been going to stay the night at Suffolk, so wash gear, and a change of clothes are in the boot. In addition to the practical and very fine Jaguar toolkit, I have my own tools on board, plus a tape recorder to prattle to. In the brief call at my empty home, a case of cassettes was thrown in; the XJ6 is one of the relatively few cars which are quiet enough, more than quiet enough in the Jaguar's case, for music to be enjoyed at speed. It ould help avoid drowsiness. ot that I was tired now – and in ny case I have a rule, reliably nforced by sleep and one hint ng ago of my eyes shutting uring a long overnight drive, at directly I feel drowsy, I stop nd sleep, wherever and as near s possible wherever that appens. The kip usually lasts etween 10 and 20 minutes; terwards, I am fit for another vo to four hours.

larvellous ride

There is something very fitting out spending my last 12 hours so with the car driving it. Time recall its virtues, so many of em, and its very few eaknesses; time to remember hat it has done for me, and for lleagues.

Its ride is a marvellous hievement. You cannot say it is ft – yet it is virtually never oubled, except at very low eeds by a very sharp bump, nich few cars will cope with. It's its best when driven quickly er a bumpy road. You can feel e wheels pattering underneath, t never getting into an damped frenzy. This is as true w, at close to the 30,000 miles nich damper makers tell you is e time to renew, as at any evious mileage. That classically w-camber-change, all-shbone-geometry suspension es a marvellous job in sorbing unevennesses with the ast flap. The only sort of bump at can catch it out momentarily a major lifting one – a long w frequency ridge where the r's gracefully long 112.8in. neelbase causes it to flop over e top, as the nose tries to ntinue upwards on the falling de of the hump. Citroen's CX es the same but much more ughly.

The fuel is low, so we pull into 4's London-end service area at ston. Brimming the two minally 10 gallon tanks in each ar wing is always a bore, with o instead of the usual one of ose irritating anti-expansion erflow space systems to cope th. It is possible to persuade up 21.6 gallons into the XJ6 nkage and pipery, if you are prepared to spend a total of around ten extra minutes topping up until the last bubble has come up each of the fillers.

On leaving I instinctively remember the advice of Sid Creamer, of R. A. Creamer and Son, the Jaguar specialists in West London who have looked after HRW 287V so well, and switch over to the left hand tank. It's not necessary this time, because, all being well, I don't plan to leave the car standing for any time with the ignition off and the left tank still untapped; if one does, the electric feed and return changover valves relapse when de-energising to the left, and can cause some fuel to be shunted back to that tank, with consequent overflow.

Another quirk that has become habitual after checking oil levels or anything else under the bonnet, is to give the rear corners of the bonnet a good push downwards. It has been dealt with once, but the fault has returned – the left catch releases during a journey so that the rear corner of the bonnet lifts slightly, exposing its untidy return.

We set off westwards at the so easy cruising speeds with which this car, even more than the vast majority of modern machines, demonstrates so effortlessly. Accelerating does produce some engine noise – a low sound, a sort of smooth beat, muffled but noticeable, but not much more. On part-throttle cruising, which for this Jaguar can mean up to a nonchalant 110 mph, the engine's voice dies away to a murmur. The only other time one is conscious of the power unit is surprising, on tickover, when there is an occasional little shudder, which you can feel, but which doesn't show on the revcounter.

Little noise

There is a little road noise, the otherwise superb Pirelli P5s making one pay a small price for their delightful response and adhesion. Wind noise is present too, not much by normal standards, but a little more than I recall on previous Series 2 Jaguars. Autotest experience has taught that wind noise has little to do with body shape in cars, and far more with how good door seals are, but I cannot help wondering if the angularities of the Series 3 Pininfarina changes are as quiet as the aesthetically perfect curves of the Lyons original.

For one of the reasons why the XJ6 body has always pleased my eye so very much is its sleekness. One remembers again the suggestion that, as first launched in 1968, the car had an element of the sports saloon in its conception. It is still there, in the Jaguar's lower overall height (54in. to the around 56½in. of its BMW and Mercedes rivals), and slightly narrower build (69.7in. wide to 70.9 for the 735i and 71.7 for the 500SE) which combine to give it between 1½ and 2 sq ft (5.8 and 7.3 per cent) less frontal area. One feels that one is pleasantly nearer the road – without being unpleasantly close to it – so that even lesser large saloons like Ford Granadas look (and are) high.

Midnight sees us at the Membury service area, reclining the seat for a three-quarter hour snooze. Awaking as the interior cools, I reset the seat rake, once again thinking that such a very well equipped car should really have stepless rake adjustment, ideally combined with coarse adjustment as on Rolls-Royces, instead of the notched lever arrangement. However, it does have a position that, with the help of the lumbar support adjustment, has suited me well enough over the last year. I re-start the engine – how perfectly the Lucas-Bosch L-Jetronic injection makes this great power unit start, *always* reliably, and always without fuss or hesitation. I remember ancient prejudices against fuel injection and its risky-seeming complexity compared with carburettors, how one doubted its long term reliability, and how reliable the Bosch systems seem to be today. Settling into the seat, relaxed in the ideally placed (for me at any rate) door and centre elbow rests – something I miss in other cars – I set off again, fresh and ready for as many miles as the rest of the night is long. It is that sort of car; it invites, encourages you to drive, with its perfect manners and what a friend once fittingly called "obedience".

Revelling in the delight of a deserted M5 past Bristol at around 1.20 am, I set the AE cruise control, something that British motoring rarely allows because of too much traffic. It is a good one, maintaining a steady speed with much more subtle corrections than the Bosch one used by BMW, and comparatively smooth in its pick-up when you re-engage it after cancelling it with a dab of the brakes.

The lazy way

Cruise control is decadent motoring, giving the lazy satyr at the wheel a flattering feeling of indolent power. Having caught up with a slower vehicle and disengaged the control the refined way (covering the accelerator with one foot to cut out the small lift-off jerk when gently braking), and inspected the other driver, he simply dangles the left hand languidly over the centre armrest to touch the resume switch, and the car surges gently but firmly forward again back up to its cruising speed.

Entering Somerset at 1.41 am the emptiness of the motorway allows one to enjoy the very fine headlamp system on full beam, when it has both great range and a wide spread of strong white light, making the cats' eyes into an endless necklace of bright diamonds which stream past one in the road, and rendering the faint concrete of distant bridges firm and hard as they approach and sweep overhead. The extraordinarily vile smell of whatever it is near unfortunate Bridgewater brings back memories of draughtier night drives in a Dellow on the MCC Lands End trial. Soon after the Devon border, although feeling still perfectly awake, I wonder about hallucinations, since we suddenly find ourselves running past a collection of well-lit, empty, curtainless rooms, in a long single storey, themselves moving in the same direction nearly as fast as we are. Alongside

That still so handsome engine; note screw oil filler cap, which gets very hot and would be better à la Rolls Royce, with a snap action release

The right way to enter your 30,000 miles, at a true 135.5 mph (the car's downwind best speed – its mean maximum in imperfect, windy conditions was 129.8 mph)

WHAT THE JAGUAR XJ6 HAS COST

Total for 12,000 miles £8,282.55. **Cost per mile** 69p

Mileage now: 30,000. Age now: 14 months

Period covered by this report: past 12,000 miles and 6 months

**Based on use over 12 months and inc. standing charges*

CONSUMABLE ITEMS	Life in Miles	Unit cost	Cost per 12,000 miles
Fuel: 4-star (gallon)	17.4	1.30	896.55
Oil: topping-up between changes (litre)	2,640	1.85	8.41
Brakes:			
Front pads (unit cost = set of 4)	24,000	32.20	16.10
Rear linings (unit cost = set of 4 on exchange)	15,000	16.10	12.88
Tyres: Pirelli P5 205/70 VR15in.			
Front pair	27,000		107.20
Rear pair	29,000		99.81
(Unit cost = single tyre)		120.60	

SERVICE AND REPAIRS		
Recommended charges for service at £10.00 per hour (labour only)	£83.87 at 6,000 miles £136 at 12,000 miles	
Service costs incurred with our car in past 12,000 miles (inc. oil and materials)		£326.72
Repair costs incurred		
Total maintenance costs incurred		£326.72

STANDING CHARGES	
Insurance (see note) for 12 months	£154.88
Tax for 12 months	£60.00
Depreciation estimate	
Cost of our car when new	£14,609
Value at previous 12,000-mile report	£10,500
Value today (car 14 months old)	£10,000
Estimated depreciation (12 months)	£6,600

SUMMARY	
Total running costs (consumable items plus Service and Repairs)	£1,467.67
Standing Charges (12 months)	£6,814.88

Note: *To put all our cars on equal footing for insurance cost, the figure given above is a typical quotation for a "good risk" driver – with clean record, and car garaged in Oxfordshire, a "middle range" risk area. Full n.c.b. discount has been deducted, as has the saving for £25 excess. The actual figure given is the middle one of five quotations ranging from £150.48 to £169.60. Source: Quotel Motor Insurance Service.*

MAXIMUM SPEEDS

Gear	LT mph	LT rpm	RT mph	RT rpm
Top (mean)	**130**	**5,250**	127	5,150
(best)	**135**	**5,500**	130	5,250
2nd	**85**	**5,000**	85	5,000
1st	**51**	**5,000**	51	5,000

FUEL CONSUMPTION

Overall mpg:
LT 17.4 (16.2 litres/100km)
RT 16.8 (16.9 litres/100km)

Note: *"RT" denotes performance figures for XJ4.2 automatic tested in* Autocar *of 29 December, 1979.*

Standing ¼-mile:
LT 17.4 sec 85 mph RT 17.4 sec 85 mph

Standing km:
LT 31.8 sec 102 mph RT 31.6 sec 103 mph

ACCELERATION

FROM REST

True mph	Speedo mph LT	Time secs LT	Time secs RT
30	**31**	**3.8**	3.8
40	**41**	**5.7**	5.6
50	**51**	**7.5**	7.4
60	**61**	**10.1**	10.0
70	**72**	**13.3**	13.3
80	**82**	**17.1**	16.9
90	**92**	**22.1**	21.9
100	**102**	**29.7**	29.4
110	**112**	**38.9**	38.5
120	**122**	–	–

IN EACH GEAR

mph	Top LT	Top RT	2nd LT	2nd RT	1st LT	1st RT
10-30	–	–	–	–	**2.9**	2.9
20-40	–	–	–	–	**3.4**	3.3
30-50	–	–	**5.3**	5.3	**3.3**	3.3
40-60	–	–	**5.6**	5.5	–	–
50-70	–	–	**6.0**	5.9	–	–
60-80	–	–	**6.8**	6.7	–	–
70-90	**10.8**	–	**10.7**	–	–	–
80-100	**12.7**	–	**12.5**	–	–	–
90-110	**16.4**	–	**16.2**	–	–	–
100-120	–	–	–	–	–	–

SPECIFICATION

ENGINE	Front, rear drive
Cylinders	6, in line
Main bearings	7
Cooling	Water
Fan	Viscous/Electric
Bore, mm (in.)	92.07 (3.62)
Stroke, mm (in.)	106.00 (4.17)
Capacity, cc (in³)	4,235 (258)
Valve gear	Dohc
Compression ratio	8.7-to-1
Fuel Injection	Lucas/Bosch L-Jetronic
Max power	205 bhp (DIN) at 5,000 rpm
Max torque	236 lb ft at 3,700 rpm

TRANSMISSION

Gear	Ratio	mph/1,000rpm
Top	1.0-2.0	24.7
2nd	1.45-2.90	17.1
1st	2.39-3.78	10.3
Final drive gear	Hypoid bevel	
Ratio	3.07	

SUSPENSION	
Front – location	Independent, wishbone
springs	Coil
dampers	Telescopic
anti-roll bar	Standard
Rear – location	Independent, lower wishbones, fixed length drive shafts
springs	Twin coil
dampers	Twin telescopic
anti-roll bar	Standard
STEERING	
Type	Rack and pinion
Power assistance	Standard
Wheel diameter	15¾ in.
BRAKES	
Front	11.2in. dia. disc
Rear	10.4in. dia. disc
Servo	Vacuum

PRODUCED BY:
Jaguar Cars
Browns Lane, Allesley,
Coventry
Warwickshire (CV5 9DR)

motorways, trains are sometimes surprising sights.

The end of M5 at junction 31 beyond Exeter at 2.21, with 193 miles on the trip, and 29,503 on the main recorder, puts me briefly on to a great twin track near-motorway road they say is A30, which seems wrong when I remember the narrow meander which A30 used to be not so long ago. We turn round at the junction with A377 to Crediton and head back for M5 and the north-western leg.

The first fuel stop is at the Gordano service area before Bristol at 3.51, which shows that the car has done a true (corrected) 17.1 mpg since Heston. This figure is typical of good long fast runs away from traffic, representing the poorer end of the better consumption part of the Jaguar's fuel useage spectrum. I recall our idyllic summer holiday in Connemara, when in relaxed mood, avoiding the temptations of the throttle pedal and more or less observing Ireland's 60 mph overall limit, the car returned 20.9 mpg corrected overall, which included nine days of summer-cold starts and some town running in Dublin. This was not a flash reading; I have done likewise on other occasions. But its normal figure remains at around the high 16s and low 17s, dropping to 15 or so in any great deal of cut and thrust town running, short runs and truly cold starts.

Stability

Rain forecast for the early morning appears as very fine spray on the screen on leaving Gordano and is the real thing five minutes later as we cross the M4 heading Birmingham-wards with 325 miles done. The car's straight stability is another reason for its superb cruising abilities. It had begun to move very slightly from the straight a little earlier, and I hadn't been surprised on getting out for the fuel stop to find that a fair side wind had sprung up. Th odd vehicle now shares the motorway with me, mostly lorries heading the same way, and I again appreciate another o those little details that make a great motor car nearer than mos to perfect – the way the dip switch stalk works, with no soun or feeling of a click as you pull it back and let go, Continental styl

The 400 mile mark appears as we pass Birmingham at 5.07, together with the decision not to turn round and return via M4 bu to continue north, then east and south on the M6 Midlands Link and M1 back to London. The slowing for the M6 eastward junction shows up one minor momentary failing of all Jaguar XJ brakes, when after a long we run they tend to pull slightly one way or another during the first part of the stop; they quickly so themselves out to pull evenly, b in such a stable car the first symptoms are a little disconcerting.

M6 at 5.12 is quite busy, and

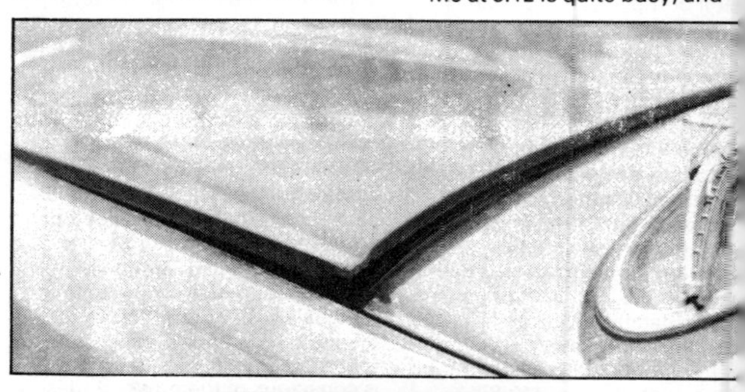

Bonnet catch on left latterly tende to release partay during a journey

when departing after a 40 minut snooze at Cawley service area near Coventry at 6.06 (428 miles) I am struck by the tremendous contrast with M4 five and a half hours earlier; the motorway is relatively crowded with cars and lorries to an amazing extent. M1, joined near Watford Gap at 6.?7 is the same, so that the drive south after a 21.0 gallon fuel stop (17.2 mpg) is not much different from what it will be like for the rest of the day. I remember the long queues one sees trying to join London's North Circular Road at the Cricklewood end of M1 at this hour when we are on our way to MIRA road testing, bu mercifully at 7.16 there is no bother.

The same cannot be said for the North Circular going wes, and we quickly tire of stop-start crawl after some decent motoring, and fiddle our way left

›nd south via Paddington on to he A4. The final homecoming un along the wide and only niddling-crowded Cromwell Road extension is glorious, to the ull blast of Tchaikovsky's *Marche Slav* from Radio 3, and the Jaguar using its power and priceless response to ease from ane to lane, making the most of raffic opportunities without ntagonising anyone. My home oad is entered at 8.06 am, nine hours and 537 miles after our departure, and I listen for what vill probably be the last time to he squeak from the power steering on full lock as I turn the ar round.

There is time for a quick breakfast, before knocking the inal 153 miles off to that ndicated 30,000. Forty minutes ater, I am turning the key in the driver's door and listening to the clunk of the central locking system. Driving away, I note that he speedometer needle has a iny shake in it, suggesting a cable strand has broken, and that is before the change quality of he Borg Warner Model 66 varies lightly, with exemplary smoothness most of the time, but occasional lapses into a noticible jerk on some full throttle 5,000 rpm shifts from second to top.

The little run back up M1 is uneventful, and much the most enjoyable part is after leaving the motorway for A5 from the A45 crossing near Weedon. Motorways divorce driving from corners, which the Jaguar copes with truly marvellously by any standards — not much roll for a big saloon (that low build again, plus other stronger reasons of course), not a great deal of understeer, and very good grip. Nearing the testing site, I do mental arithmetic to work out how many miles we must do in order to have the petty satisfaction of recording what the speedometer and revcounter say t the car's maximum speed when the total distance recorder is showing 30,000 miles.

Sums and the necessary mileage done, the car obliges, in spite of the wind, with a mean top speed of 129.83 mph, and a best one way of 135.50, which as the camera shows, corresponds to an indicated 139½ and 5,550 rpm — the last figure compares with a calculated 5,500 rpm red line. Any one who has been testing XK-engined Jaguars for any time will confirm that the makers will unofficially tell you that such a crankshaft speed is tolerable for short periods; it is in fact only relatively lately that the red line has been brought down from 5,500 rpm. The rest of the performance checks show that HRW 287V is just as fast as ever, and so we set off for Browns Lane feeling that, apart from a by now badly needed wash, honour is satisfied.

Hard life

Mechanically, this Series 3 XJ6 4.2 is in marvellous health. Having said that, it should be remembered that the car has had an untypically hard life. It began its career with us as the Autotest car (*Autocar* 29 December 1979) which means a lot of very heavy use. It has been driven in all sorts of conditions, often hard. Its performance has been checked three times, at 2,750 miles, 17,000 and 30,000 miles, which means tests of flat-out acceleration and top speed on all occasions and of ultimate cornering, braking and ride at least once.

Up to the time of the first 12,000 mile report (at 17,000 miles, *Autocar* 11 October 1979), the original electric arial had played up (or rather stayed up partially) and been replaced with a far better, faster-erecting and entirely reliable Merlin type; air conditioning and fan-cum-water and steering pump belts had to be replaced at 12,000 miles; a small but just discernible click on accelerate/over-run from the final drive appeared at 15,000 miles but has got no worse in the subsequent 15,000; central locking misbehaved for a while between 13,000 and 15,000 miles when it was dealt with; the lid of the glove locker became stiff and needed re-setting at the same time; the small yellow plastic cap for the huge 1.6-gallon washer reservoir (so big that it usually lasts without refill for 5,000 miles of winter driving) cracked and regularly fell off — its replacement at 24,000 miles rapidly cracked and behaved likewise; paint tended to chip too easily, a result almost certainly of the paint plant problems Jaguar suffered from in 1979, but which are reported to be cured for some time now.

As reported in the last long term article, the Pirelli P5s add to their handling virtues the satisfaction for the owner of good life. If one had allowed them to wear down to within a millimetre of the legal 1mm limit, I am confident that they would have achieved 27,000 and 29,000 miles as predicted in that report. But their wet grip was beginning to drop off slightly at 25,000 miles, just before my colleagues Tony Howard and Peter Cramer were to take the car as transport to the Paris show, so I played safe and fitted a new set, they have not had to put up with the rigours of Road Test mileage.

It is a jewel of a car. It has overall no equal internationally; there are faster big saloons — but either they do not handle so well, or have not the same roadholding. None of those presently made are as quiet — and although the Jaguar price list today is not so highly competitive as it used to be (did the makers foolishly take note of what so many motoring journalists glibly said about pricing Jaguars more highly?), none of the others can be bought for quite as little money. Since the coming of the Series 3, Jaguar have lengthened their minimum service interval from that irritating 3,000 miles to 7,500 miles, which usefully reduces running costs. The replacement car is still some way off. Contrary to what some write, the present range is not dated; it is in many ways timeless. But a new, lighter, more fuel-efficient car must come. The job of making such a Jaguar, which must be at least as quiet as the present XJ saloons, must be very difficult. I hope Browns Lane manage it — and I have the feeling that they are the only design team who could manage such an achievement. □

Headlamp wash/wipe works well on larger lamp, but is not provided on smaller one

Below: A pleasing Jaguar detail: air conditioning pump has threaded stay to make belt adjustment easier, as has generator

Bottom: Bumpers have black moulded outer surfaces, to protect from minor scratches

Very combined lighting switch annoys strangers to the car with its complexity, but owners get used to it

Remote controls for electric door mirrors had started to crack their rubber towards 20,000 miles, but this did not affect their working

CAT WITH A LONG TIN ROOF

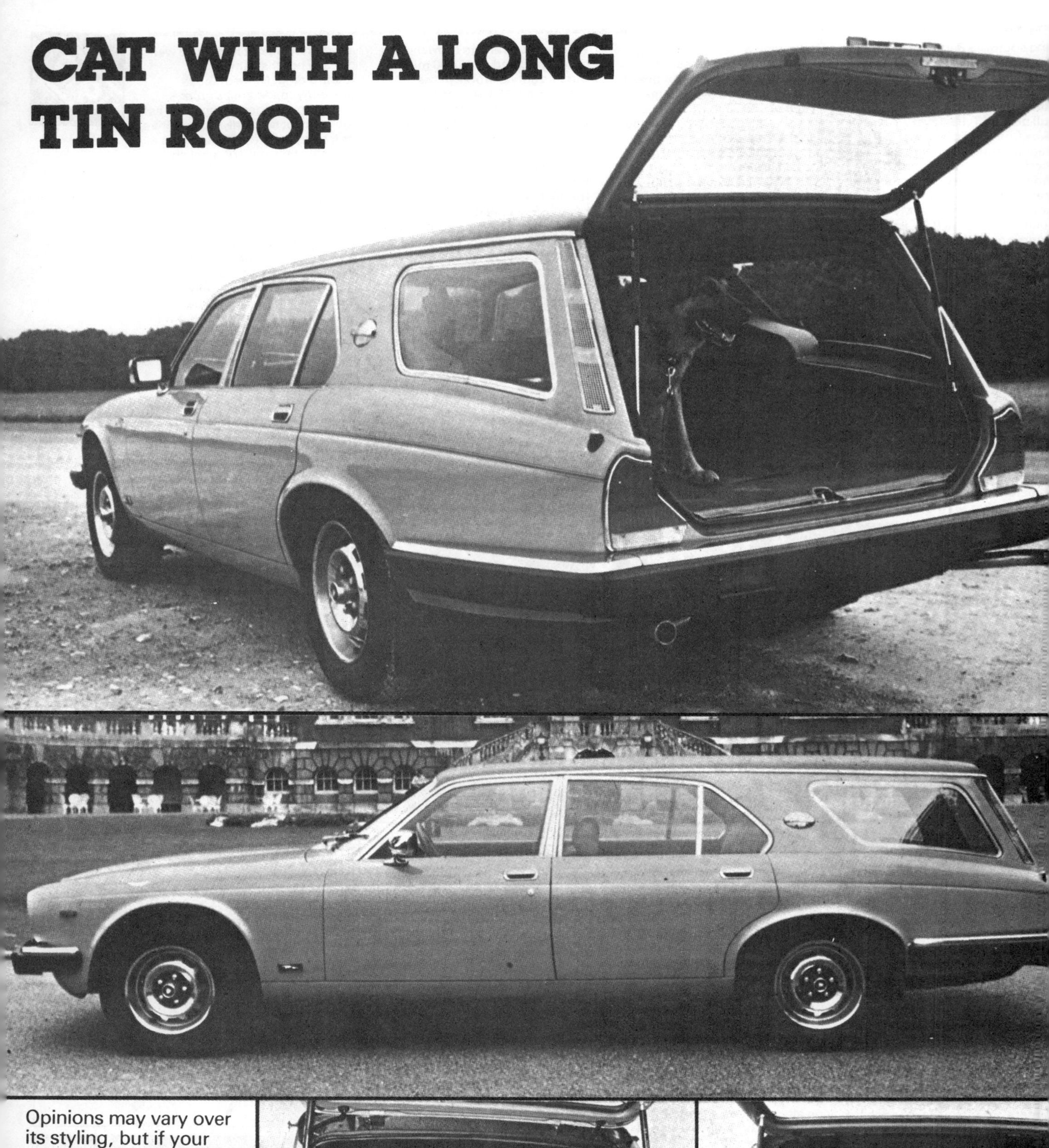

Opinions may vary over its styling, but if your antique Chippendale simply refuses to be transported from A to B in anything less than XJ opulence, there's only one car for the job: the Ladbroke-Avon XJ Estate. Jeremy Sinek takes up the story overleaf. Photographs by Peter Burn

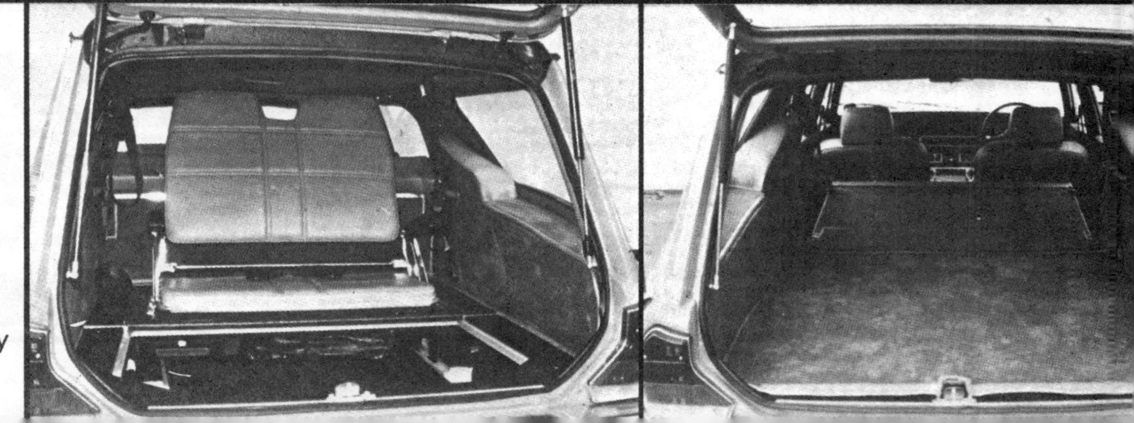

PERHAPS THE most surprising thing about the idea of a Jaguar estate car is that no-one has ever done it before. After all, in the lack of any suitable offerings from the original manufacturers of prestige saloons, there is plenty of precedence in the form of specialist coachbuilders offering their own estate car conversions on such cars as Rovers and Mercedes. Perhaps the main stumbling block was the difficulty of reconciling the sensuous curves of Jaguar styling with the inevitable boxiness of an estate rear end.

In the end it took the Warwick firm of Ladbroke Avon Ltd (see Philip Turner's story elsewhere in this issue) to take the cat by the tail by removing the top of the bodywork from the B-post rearwards, and replacing it with a new roof extending all the way back to an opening vertical tailgate. The result is hardly a paragon of pulchritude, but then again it is hard to see how it could have been done a great deal better without major surgery to the rear door frames, which would have added even more to the £6,500 such a conversion will cost you on top of the price of a new SIII Jaguar or Daimler saloon.

But styling apart, it is not hard to see how the Avon Jaguar Estate car won a Gold Medal for coachwork on the occasion of its debut at the 1980 Birmingham Motor Show. The interior opulence that is the hallmark of any Jaguar/Daimler is immaculately carried through to the rear, and quite apart from the fact that 'you cannot see the join', the construction of the coachwork shows some cunning touches: the way that the sweep of the rear wings has been maintained, as have the rear light clusters; and the vertical part of the boot lid which has been subtly grafted onto the upper part of a tailgate purloined from, of all things, a Renault 5.

Any assessment of such a conversion must, however, raise two fundamental questions — how practical and capacious is it as an estate car; and how good is it still as a Jaguar?

After all, there was a time when estate cars were synonymous with noisy utilitarianism; were the conversion to reduce Jaguar's legendary comfort and refinement to the levels of more mundane machinery, the whole exercise would be self-defeating.

According to Ladbroke-Avon the Estate is heavier than the equivalent saloon by an insignificant 40 lbs so there should be little loss of performance. In fact, subjective impressions were that the Estate was if anything *more* lively than the SIII 4.2 saloon we tested in 1979, but that is probably due to differences in gearing (the Estate was on the standard 3.31 final drive, the saloon on the optional — and preferable — 3.07) and the fact that the Estate was more willing to kick down to 1st at low speeds and to hold 1st to higher speeds under hard acceleration.

Test track figures showed the differences to be negligible on acceleration, with the lower-geared Estate losing out on top speed. Strict observance of the 5000 rpm red line would limit this to a modest 115 mph, but there is little doubt that if you cared to over-rev it, 120 mph plus would be easily attainable, if totally academic.

Despite its lower gearing, and being used for an unusual number of short journeys with cold starts, our Estate test car marginally improved on the saloon's fuel consumption with an overall of exactly 16.0 mpg — respectable enough for the class of car.

At high cruising speeds, the subjective impression is of higher noise levels than in the saloon, though it is hard to be sure how much of this is due to the lower gearing, and how much to greater levels of wind noise — which may or may not be a side-effect of the conversion. At very low speed there is no doubt that exhaust noise *is* a little more noticeable than in the saloon — reflected in a 30 mph dBA reading of 58.5 instead of 56 — but at other speeds our test track figures show only a negligible difference, which means that overall the Ladbroke-Avon Estate still numbers among the quietest cars in the world.

Tailgate is half XJ saloon bootlid and half Renault 5 rear hatch, cunningly joined

We could detect no significant loss of structural rigidity, nor any difference in the fine combination of ride and handling which numbers among the many other XJ virtues — qualities which still add up to one of the most soothing cars in the world in which to drive or ride.

As an estate car the Jaguar is undoubtedly a useful device, if not as commodious in every direction as some lesser alternatives from Volvo and Citroën, for example. With the rear seat erect, there is a very regular-shaped and flat rear cargo deck which measures exactly 44 inches square. There is no loading sill, and the rear aperture is roughly 41 inches square, the width largely dictated by the position of the supporting gas struts, which are also responsible for the tailgate not lifting as high as one would like to avoid the risk of dented foreheads.

For maximum capacity the rear seat backrest can simply be tipped forward onto the fixed cushion; it does not lie completely flat, but gives an effective floor length of 75 inches, or a few more if you allow the contents to overhang the leading edge, the ultimate length depending on where you set the front seats. With plush carpeting and leather trim, however, there is little incentive to carry loads that are less than spotlessly clean.

An extra fitted to our test car was a rearward facing child's bench seat, which disappears completely under the 'boot' floor when not in use (though at the expense of a useful amount of hidden underfloor storage space), and which greatly obscures rearward vision when erect. It costs £475.

We are sure that Ladbroke-Avon themselves would agree that if you truly need the ultimate in load-lugging capacity, there are roomier and more practical estate cars than this Jaguar. Then again, there's not a great deal wrong with it either as large estate cars go. More to the point, however, is that if you *need* an estate car, but you *want* that special brand of motoring that only a Jaguar or Daimler can provide, what else is there?

When you put it like that, Ladbroke-Avon's plan to build no more than 250 of the machines over the next couple of years seems likely to leave a lot of disappointed would-be members of what promises to be a very exclusive set.

PERFORMANCE

	Estate	Saloon
Maximum speed	120.0*	128.0*
*see text		

ACCELERATION FROM REST

mph	sec	sec
0-30	4.1	4.2
0-40	5.8	5.9
0-50	7.9	7.7
0-60	10.5	10.5
0-70	13.7	13.5
0-80	17.2	16.9
0-90	22.3	21.6
0-100	28.7	28.4
Stand'g ¼ mile	18.0	17.6
Stand'g kilometre	32.2	31.8

ACCELERATION IN KICKDOWN

mph	sec	sec
20-40	3.4	3.5
30-50	3.8	4.1
40-60	4.7	5.5
50-70	5.8	5.7
60-80	6.7	6.0
70-90	8.6	8.2
80-100	11.5	11.8

FUEL CONSUMPTION

	mpg	mpg
Overall	16.0	15.7

NOISE

mph	dBA	dBA
30	58.5	56
50	64.0	63
70	69.5	69
Max revs in 1st	72.5	72

Maker: Ladbroke Avon Ltd, Ladbroke House, Millers Road, Warwick CV34 5AP. Tel: Warwick 41377/8/9

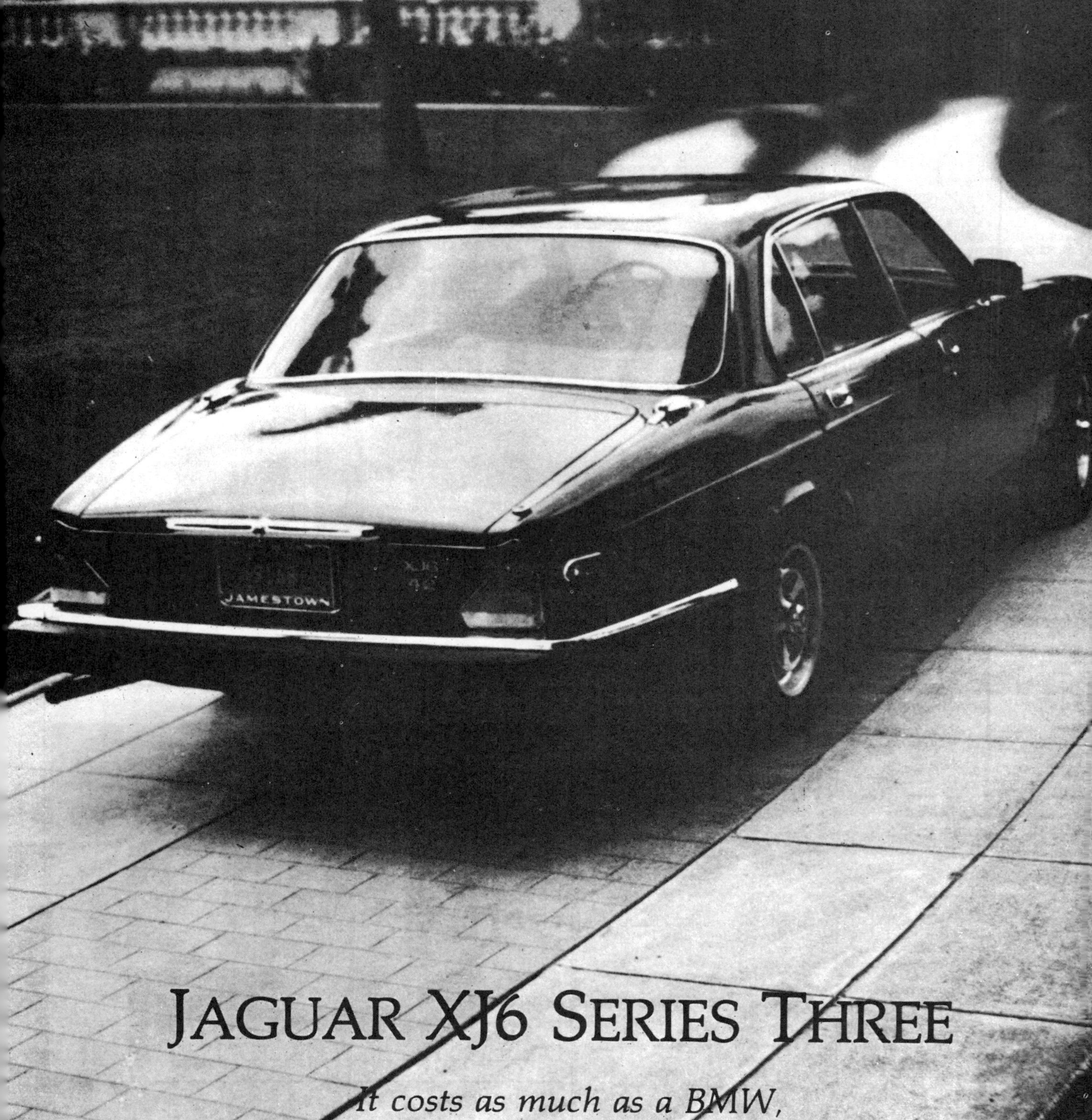

JAGUAR XJ6 SERIES THREE

It costs as much as a BMW, but it is really a bargain-basement Rolls Royce.

By Jim Hall

HERITAGE. There aren't many cars you can buy in this country that have it. History, yes, heritage definitely not.

The Jaguar XJ6 is steeped in heritage. When you look at a Jag sedan of the past be it a Mark VI or a 3.8S or even a 420G you know it is a Jaguar. And it's not just a radiator grille you identify, its something about the whole car and may go quite beyond the visuals.

The XJ6 has been with us for nigh onto thirteen years now. A few may remember the car was introduced as the replacement for the Jag 420 and Daimler Soverign models (Daimlers are "badge-engineered" Jaguars with fluted radiators) in 1968. While two engine displacements were offered, the U.S. market saw only the larger 4.2 liter XJ6. While fine road cars, early XJ6's were a bit difficult to keep on the road. Yet the new Jag sedan proved to be quite successful, so much so the model range was greatly expanded in subsequent years.

1973 saw the introduction of the XJ6L or long wheelbase model. This was essentially an XJ6 Deluxe Saloon with an additional four inches spliced into the wheelbase. The following year brought the XJ6 Series Two. A minor restyling job was done on the nose of the car to visually lower the grille and to accommodate U.S. Government-mandated impact resistant bumpers. '74 was also the year Jag introduced the beautiful XJ6C coupe. These were short wheelbase XJs sporting a pillarless hardtop two-door body. Leave it to somebody to muck-up a car this gorgeous. Leyland fitted them with vinyl roof coverings. Yuk! Well at least it could be stripped off.

In 1975 the short wheelbase four-door was discontinued thereby stratifying the range with two-door coupe riding on a 108.8 inch wheelbase and all four doors running with 112.8 inches 'twixt their front and rear axles. The coupes survived until the last third of the 1977 model year when they finally went the way of the original XJ short wheelbase sedan. Then during the 1980 model year Jaguar introduced the XJ6 Series Three. With the Three all Jaguars for all markets were fitted with U.S. style bumpers. The radiator grilles were redone yet again, door handles became flush with the bodywork. New rear lamp clusters were fitted and the entire upper of the car redesigned. This last change is the most significant, for the new roof afforded rear seat passengers further head room, did away with the fixed quarter lights in the front doors and eliminated the chrome clips that were used to help retain the rear screen at speed. It also made a tremendous improvement in the cars appearance without calling attention to the change. In fact, most observers are not aware there has been a major sheet metal change made on the Series Three. While the 1981 XJ6 differs very little from last year's model, the newest Jag sedan is, overall, the nicest XJ yet.

Remember heritage. On the XJ6 the mechanicals simply reek of it. The engine, a fuel injected 4.2 liter version of the XK Series powerplant will go down in the history books as one of the longest continuously produced powerplants in automotive history. Jaguar began building them in 1948. To put that in perspective, that was a full seven years before Chevy started putting small block V-8s together. It's even more impressive when you look at the engines configuration; dual overhead camshafts, hemispherical combustion chambers in a light alloy head and Bosch fuel injection. To be sure it was advanced when first introduced but the XK's specifications are still anything but common. This is not to say the engine hasn't started to show its age. Emissions controls have sapped quite a bit of the XK's once legendary horse-

power. The specific power of the engine is considerably below that of a few sixes that power other European luxury cars. The big 4.2 liter powerplant is quite thirsty in addition. The average economy the XJ6 turned in for the total test period was a bit over 17 miles per gallon.

There are a number of plusses that outweigh the engine's minuses. This is one smooth six, quiet too. In addition, in spite of being down on horsepower when compared to a few of its German competitors, the car does move along quite well, thank you. Yes, it is a bit slow off the line but once rolling the car does zip. This is to be expected, I suppose, in light of the XJ6's horsepower rating and 4100 pound plus weight.

The big Jag's handling is very, very good. Basically an understeerer, the XJ is predictable, and direct. The line will tighten slightly if you back off the throttle in mid corner but nothing radical is going to happen. The rack and pinion steering is up to the rest of the chassis with just the right amount of power assist. About the only complaint one could make about the steering is the 3.3 turns lock to lock of the unit would be nicer if quickened up to about 2.8 turns. But perhaps I pick nits.

Inside the XJ6 is everything you expect from a Jag. Connoly leather fills the interior with an aroma that is equalled by only one other car. The instrument panel is burr walnut and one of those things that makes a Jaguar a Jaguar. The instruments are logically layed out and would be most legible if it were not for their flat lenses that do cause quite a problem with reflections when driving with the sunroof open. The primary gripe I've got with the interior is the steering wheel. The rim is just too thin for my tastes and the

center pad is just not up to the level of the remainder of the design.

The Series Three Jag is every bit a Jaguar. Another of the breed to carry on the tradition. One part of that tradition has been the previously mentioned unreliability. I put this in the past tense on the comments of our man in England, Alan Clark. Clark relates that Jag reliability has improved dramatically over the last three years. I have to add parenthetically that Jaguars in Britain seem to have a better frequency of repair record than on this side of the pond. Don't ask my why, I'm not sure but I suspect the folks in the British Isles may respect the heritage of the cars mechanicals and take care of them better than we.

With reliability improved as well as engine drivability, thanks in no small part to the fitting of fuel injection, the Jaguar XJ6 Series Three sedan is one of the markets current good deals even at $27,500.00. Jag likes to think of the XJ as BMW Seven Series competition or perhaps the thing to take Mercedes' 280 sedan on head to head. I think the car will give both of the aforementioned cars a run but Jaguar have really aimed low. You see, I'm gonna tell you a secret: the XJ6 with climate control air conditioning, cruise control, central locking sunroof et al is every bit as good as a Rolls Silver Spirit in my book. True, it doesn't have the Roller's resale, radiator, degree of hand finishing or snob appeal. But it doesn't have the Spirit's weird hydraulic system, Chevy Caprice looks or home sized price tag either. However, the XJ6 will drive circles around the Rolls, gets better (although still not outstanding) fuel economy, has the same Connoly and walnut interior with better instrumentation all in a package as sexy as Kate Bush. No Jag missed the market. It costs as much as a BMW but it is really a bargain basement Rolls Royce.

And that, in fact, may be the part of the illusive Jaguar heritage you can't see.

To me, it's the most important part.

PHOTOGRAPHY: AARON KILEY

Jaguar XJ6

Classic motoring for sale. Inquire within.

• Automobiles change. Standards do not. Only one car has endured as the standard of measure for luxury sedans while Mercedes have become futuristic, BMWs have grown soft, and Rolls-Royces have been downsized: the Jaguar XJ6 is still the Jaguar XJ6. It is that rarest of modern automotive quantities, a classic, and you can pin down exactly what makes it so.

The car's classic status begins with the materials used in the interior. Leather adorns the seats and the door panels, while the instrument panel features walnut burl and unpretentious Smiths instruments. These materials make you feel good because, like the English countryside, they are both civilized and earthy at the same time. When people order leather interiors for their new cars, they have an image of the Jaguar XJ6 in their minds. But nowhere except in the real thing does this image fully come to life.

The engine is also a classic, basically an updated version of the straight six first fitted to the 1948 Jaguar XK120 sports car. Cylinder-head wizard Harry Weslake was hired by Jaguar to update its Thirties-vintage, long-stroke six, and he flowed the head and specified hemispherical combustion chambers and a double-overhead-cam valvetrain. The result was powerful enough in prototype form to power Major Goldie Gardner's EX135 MG streamliner to record speeds, and a decade of Jaguar competition successes followed.

Jaguar's six is able to survive today because its relatively small combustion chambers permit short flame travel for complete combustion and have a low surface-to-volume ratio to minimize HC production. Also, residual emissions and drivability problems were eliminated in 1978 with the innovative mating of Bosch/Lucas L-Jetronic fuel injection with a three-way catalyst. But the main reason this engine survives is the thrill you get when you leg it. A straight six is preternaturally smooth. Its crankshaft balance is in fact superior to a V-8's. Power pulses are close enough that they overlap one another, so torque pours forth in a continuous rush. The Jaguar six makes effortless work of 90-mph running; a 350-cubic-inch Chevy V-8 feels anemic by comparison. Unfortunately, the price of the Jaguar six's astonishing 176 hp is 15-mpg fuel economy.

The XJ6's suspension backs up this smooth performance with virtuosity that is still unequaled after thirteen years. There's sufficient anti-dive geometry dialed into the front suspension to permit springs soft enough for a compliant ride without the threat that load transfer under braking will make the car sink to its knees. The fifteen-inch tires help ride quality as well. Most of the goodness, though, comes from generous wheel travel. Also, shock travel very nearly matches wheel travel at all four corners. The Jaguar does move up and down on its suspension more than we like, but it rarely loses its balance. Hurl the XJ6 into an on-ramp, an experience it relishes, and the rear end gradually drifts out until you're actually cornering the car instead of grinding the front tires into dust with prodigious understeer.

The overall result is one of the finest combinations of ride and handling you can buy. The XJ6 is a driver's car, but it never presumes to occupy your full attention, acting as the perfect filter between pavement and driver. It's as if Jaguar believes a little civilized conversation should accompany full-throttle motoring.

Jaguar adds a few things to this classic package every year—an electric sunroof and an upgraded three-speed Borg-Warner automatic transmission are among the 1981 additions—but the firm's stylists still suffer lapses in taste. The new-for-'81 wheels (from the XJ-S) appear to have been taken from a J.C. Whitney catalog, while the black seatbelts (on a tan interior) and paint-color coordinations are simply shoddy.

Quality problems also dog Jaguar. The new two-year/50,000-mile warranty on the powertrain, added to the standard twelve-month unlimited-mileage warranty on the rest of the car, indicates how hard the company is trying, but a warranty doesn't make you feel any better when something breaks. Our test car suffered a cruise-control blackout, the breakage of a power-window master switch, and an apparent oxygen-sensor failure, all before the 9000-mile mark. The radio didn't fail, but it didn't work very well, either.

Anyone who drives an XJ6 overlooks the few flaws. This car is still a classic drive, a symbol of motoring excellence. It's everything a Rolls-Royce tries to be.

But that might not be enough. Thanks to the demise of MG and Triumph, the retreat of Rover, and the fact that the 1981 model year was skipped altogether for the XJ-S, BL's corporate fortunes in America depend for the moment solely on the XJ6. Until the new higher-efficiency V-12 engine is ready for S-models in 1982, a few XJ6 sales either way could spell the difference between the continuation of a standard of measure for luxury cars in this country and the disappearance of the Jaguar marque altogether. Fortunately, Jaguar sales here are currently at their highest level in two years, so apparently a large number of Americans still appreciate classic motoring. *—Michael Jordan*

Vehicle type: front-engine, rear-wheel-drive, 5-passenger, 4-door sedan
Price as tested: $27,500 (base price: $27,500)
Engine type: 6-in-line, iron block and aluminum head, Bosch/Lucas L-Jetronic fuel injection

Displacement	258 cu in, 4231cc
Power (SAE net)	176 bhp @ 4750 rpm
Transmission	3-speed automatic
Wheelbase	112.8 in
Length	199.6 in
Curb weight	4020 lbs
Zero to 60 mph	11.7 sec
Zero to 100 mph	38.5 sec
Standing ¼-mile	18.7 sec @ 78 mph
Braking, 70–0 mph	210 ft
Roadholding, 282-ft-dia skidpad	0.70 g
Road horsepower @ 50 mph	17.5 hp
Top speed	111 mph
EPA fuel economy, city driving	15 mpg
C/D observed fuel economy	16 mpg

ON TRIAL

DAIMLER SOVEREIGN 4.2

SUBTLETY FROM THE SCEPTRED ISLE

Maybe it is only a Jaguar with "schmaltz," but the Brits can still fluke a luxury car now and then.

DON'T ENVY WEALTHY people whose accountants, or wives, tell them to dispose of some excess income on an expensive car. While the 'other half' may well live in a manner to which you'd like to become accustomed, it doesn't necessarily follow they have the world at their feet when it comes to cars.

For a start the choice of vehicles tapers off dramatically. After leaving the likes of the SDI Rover and the Statesman Caprice in the sub $30,000 bracket there are only a handful of sedans to consider. If tastes run to a sumptuous leather and wood trimmed conveyance in the traditional 'pucka' mode, then two sets of British twins will start and finish the shopping list; namely the Rolls-Royce/-Bentley duette and the Jaguar/Daimler twosome.

Of these the "Roller" and the Bentley are often thought to be too crass. Who would spend in excess of $110,000 on something vandals love and Minis out manoeuvre? Realists will look at the Leyland twins, of which the more refined version, the Daimler, has just been further enhanced in both fashion and function.

Recently we tested both the revised car and its predecessor on consecutive weeks. The model in question was the Sovereign, the bottom of the line 4.2 litre six cylinder. As the uprated car had a mere 1,000 "clicks" on the clock, the old car was used to obtain performance figures. This vehicle had seen some hard times on Leyland's press fleet, but at 21,000 kilometres was surprisingly taut and blemish free. We enjoyed it very much, but were even more impressed with machine which followed.

Externally there is but one change. Both front doors now sport remote electrically operated rear view mirrors. Inside there are several changes, while mechanically the car remains unchanged, save for the assistance given to steering. The new car has less, yes less assistance as the old system was a little overdone and felt quite 'yank tankish'.

Changes to the interior include electrically operated height adjustment on both front seats, new door trims incorporating far more effective pulls, revised map pockets and the continuation of the dash wood onto the door trim top. Minor changes have been made to the trinket tray on the centre of the dash and sundry other little tidying-up exercises such as removing the manual switch for the retracting radio antenna and making it operate in parallel with the radio on/off switch.

Additional items include rear seat head rests, personal swivelling rear seat reading lights and plush nylon rugs in the three passenger footwells. The driver, poor sod, doesn't rate a rug. Oh well, workers can't expect to get all the flow-on benefits.

What the driver does get well and truly makes up for the missing rug. The Sovereign, while far from being a tyre smoking racer, is a fast, luxurious and stable carriage. The new version is considerably quieter than its predecessor, and even at full throttle the fuel injected six and the rest of the powertrain is barely audible, unless a power operated window is opened, at which time the distinct six pot Jaguar purr can be heard.

Roadholding is excellent, the P5 Pirelli radials working well with the full independent suspension. Likewise ride comfort is hard to criticise, even on very rutted and potholed gravel roads. The car simply glides over had surfaces with nary a tinkle being heard from stones blasting into the wheel wells.

Our only road related criticism stemmed from a sprint on a gravel track. When spending near on $50,000 you would expect the car to be equipped with a limited slip differential, but for some reason Leyland have seen fit to delete this from the Sovereign only specifying it as standard equipment on the V12 Double Six Vanden Plas. While the car is quite predictable, the tail has a tendency to slide around if too much power is applied too soon coming out of a slow corner. This is great fun for boy racers, but someone in the 'golden years' may find it a bit too sur-

ABOVE: Basically, the twin overhead cam six cylinder engine dates back to 1948. During the intervening years however, it has undergone numerous engineering changes, to the extent that it could now be called "new". BELOW: Though it's neat and tidy, the cockpit still retains that traditional charm which is one of the few good things left about British cars in general. The steering wheel is adjustable for reach, and all the instruments are well placed for long term comfort.

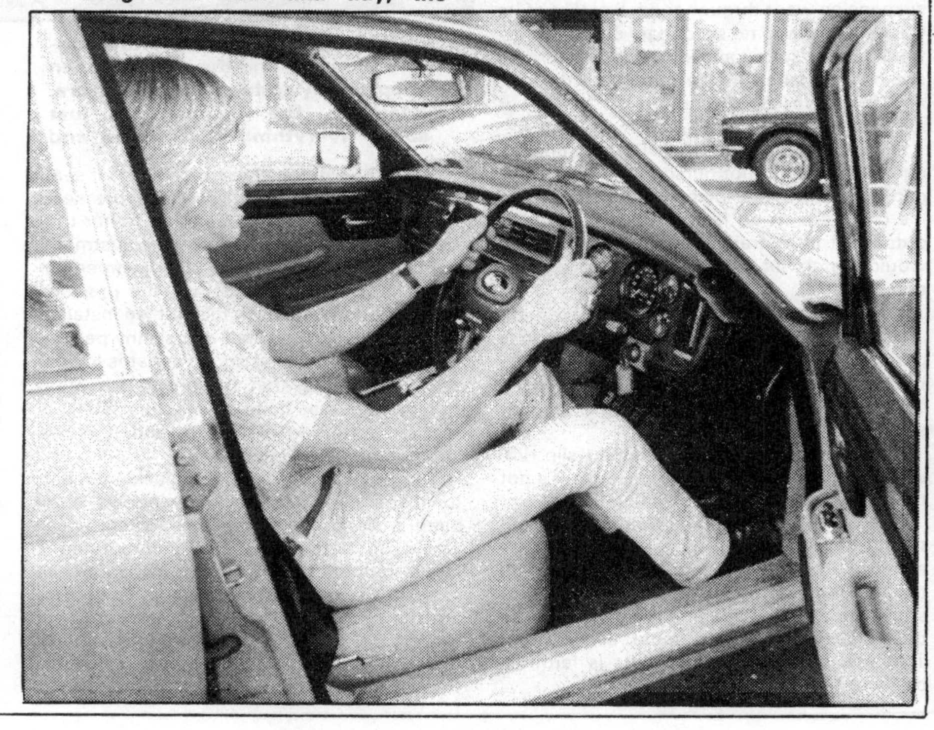

prising. Actually we overdid it on one very slow corner and manged to hang the tail into some grass where a large clump of clay was waiting to clout the sheetmetal just behind the left rear wheel!

While the clay only put a minor scuff on the paint, it succeeded in loosening one of the twin chrome exhaust extensions. Not wanting to lose what was probably an expensive part, it was slipped off the exhaust and stowed in the boot. When we arrived at our destination some fifty kilometres later we found the hot gasses had melted one of the fog lamps mounted in the back bumper. That's what you call a chain reaction!

We are not trying to shift the blame for our mishap onto Leyland by saying that a slippery diff would have prevented the problem. We would still have explored the outer limits of adhesion, but would have had a better chance of catching a slide before we blotted our record (again!).

On sealed roads the Sovereign lives up to the dictionary definition of the name; supreme and exempt from extrernal control. Pushed to the limits on tricky twisting roads, the car behaves with the calm cool headedness associated with the other definition of the word; a sovereign ruler.

It would be foolish to call the Daimler 'king of the road', but compared with many more sporty machines down the dollar scale it is quite superior.

On one particular favorite test road, which incorporates a hump-back bridge in the middle of a tight sweeper, we had the car well and truly airborn. Apart from making a five point landing with a chassis bracket hitting the bitumen, the Daimler was completely unphased by the flight and subsequent landing on a less than smooth surface. There was no bump steer, even though the suspension went through the full extent of its travel. There was no secondary change in attitude as the spring and shock absorber rates are well matched to the weight of the car, and there was no discomfort for the driver and passengers.

If the Red Brigade, IRA, PLO or finance company were in hot pursuit intent on a kidnap, then a Daimler owner could simply ask his chauffeur to go faster. Many cars could outpace it in a straight line, but few sedans would match, let alone pass it, through corners.

One final adhesion note. The brakes are superb, pulling the 1,790 kg lump up in a praisworthy 49 metres from 100 km/h. Large servo assisted discs all round are aided by anti-squat geometry in the front suspension.

Other than the lack of a limited slip diff, which we shouldn't criticise because it's not there, we have only two negative notes.

The highly polished burr walnut facia, and the flush mounted analog instruments, act as a mirror for everything to be seen through the rear window. In daylight you are able to study the clouds, birds and sundry aircraft in the reflection, while at night every street light and block of highrise flats is faithfully projected onto the dash.

TOP: The ledge on the centre console is useful, and the idea has been adopted by other manufacturers. Basically, the Daimler has just about everything that opens and shuts. ABOVE: Having two fuel tanks with their own fillers is cool! Either can be used on the move by the simple act of pressing a button.

Next, the electrically adjustable front seats come in for some criticism. Not the seats themselves, which with Connolly lether trim are extremly confortable, but the adjustment. In our opinion the range of adjustment in height is of little benefit over the previous fixed situation. To check this out we installed an array of short, fat, tall and skinny people. All ended up with seat squab set at is lowest point — proving that the previous seating arrangement was quite satisfactory. Still,it is another button to play with, and that has to be good — or does it?

The Sovereign is surely not lacking in buttons. There are those for the windows, complete with an overriding lock switch, which would be better if it also incorporated a mode to operate all four windows at once to save time when shutting up shop. Then there are the normal lights, wipers and other accessory switches, plus the mirrors, a dash map light and an interior light master switch. The interior lights feature a delayed shut down after the door has been closed to overcome fumbling for the ignition in the dark.

There is a central locking system which also secures the boot. Naturally the Sovereign has air conditioning, which we can vouch for as being one of the most efficient we've experienced, being thermostatically controlled with a variable setting control on the dash. There's also an all singing and dancing Pioneer AM/FM cassette stereo system with four speakers.

Should we be faced with a quandry spending $46,400, then there would be nothing to justify taking the Daimler Sovereign off the shopping list. In some ways it is hard to conceive as being a car. It's more like a compact lounge room with wheels. It simply glides gracefully along, no matter what the conditions, to the point where a short commuting trip in even an uptight peak hour rush is peaceful. We like it — in case you hadn't noticed!

ACTION ANALYSIS

PERFORMANCE

FUEL	ACCEL.	BRAKING
CITY 15	0-100 12.9	48
H-WAY 10.5	TOP SPEED 197	100-0

SPECIFICATIONS

MODEL...... Daimler Sovereign 4.2 Series III
COUNTRY OF ORIGIN.................... U.K.
BODY TYPE................................ Sedan
SEATING CAPACITY.......................Five
PRICE (Excluding on road costs)...... $46,400
OPTIONS FITTED...................... None
ENGINE:
Location.................................. Front
Cylinders.................................. Six
Capacity.................................. 4235 cc
Bore/stroke.......................92 x 106 mm
Block.................................. Cast Iron
Head.................................. Aluminium
Valve actuation............Twin overhead cam
Induction..............Electronic fuel injection
Compression ratio.......................... 8.7:1
Power (kW/bhp)...... 151/205 at 5,000 rpm
Torque (Nm/ft lbs)... 314/232 at 1,500 rpm
TRANSMISSION:
Driving wheels.............................. Rear
Gearbox type3 speed Borg Warner automatic
Shift location.................... Console T-bar
Gear ratios –
1st.................................. 2.390:1
2nd.................................. 1.450:1
3rd.................................. 1.000:1
Final drive ratio.......................... 3.310:1
BODY/CHASSIS:
Construction........... Unitary with front and rear auxiliary frames
Material.................................. Steel
Kerb weight.............................. 1,790 kgs
O/A length.............................. 4,959 mm
O/A width.............................. 1,770 mm
O/A height.............................. 1,377 mm
Wheelbase.............................. 2,865 mm
Front track.............................. 1,470 mm
Rear track.............................. 1,490 mm
Fuel tank capacity... 91 litres (total 2 tanks)
SUSPENSION:
Front type......... Independent, semi trailing wishbones
Springs.................................. Coil
Rear type..............Independent transervse wishbones with half shaft as upper link
Springs.............................. Twin coils
STEERING:
Type.......................... Rack and pinion
Turning circle..................... 12.2 meters7
BRAKES:
Actuation......... Hydraulic – servo assisted
Front type...................... Ventilated disc
Rear type.................................. Disc
WHEELS:
Material.................................. Steel
Diameter/width................. 15 x 6 inches
TYRES:
Make/type.......................... Pirelli radials
Dimensions.................... 205 VR70 x 15
PERFORMANCE DATA
WEATHER:................................ Hot
ROAD:.................................... Clean
ODOMETER READING............ 21,000 km
SPEEDOMETER ERROR AT
60 km/h.................................. 62
80 km/h.................................. 83
100 km/h.................................. 104

MAXIMUM SPEEDS IN GEARS:
1st.................................. 94 km/h
2nd.................................. 156 km/h
3rd.................................. 197 km/h

ACCELERATION FROM REST TO –
60 km/h..............................6.1 secs
80 km/h..............................8.8 secs
100 km/h.............................. 12.9 secs
120 km/h.............................. 17.7 secs
60 to 100 km/h.......................... 7.5 secs
(3rd gear manual or "D" auto)
400 meters.............................. 18.2 secs
Terminal speed.......................... 121 km/h

BRAKING FROM 100 km/h TO STANDSTILL
Average of four tests..............48 meters

FUEL CONSUMPTION:
Published as 2077 figures: –
City:......... 15 litres/100 kms (18.8 mpg)
Highway: 10.5 litres/100 kms (26.9 mpg)
Motor Manual figures on test: –
City:...... 18.4 litres/100 kms (15.3 mpg)
Highway: –...... litres/100 kms (– mpg)

NOTE: No special driving techniques are employed to obtain fuel consumpion figures on test. Unless otherwise stated, Motor Manual's figures are those obtainable under normal driving conditions.

SER 308

PHOTOGRAPHY: WARWICK KENT

ROAD TEST

FIVE~SPEED FLIER

You can now order your Jaguar XJ 4.2 with a manual gearbox — we pit the auto against the new five-speed and on the way to deciding which is best, take the opportunity to rediscover one of the world's finest cars.

IF THE Mercedes-Benz S-class is the WHEELS Car Of The Year and has taken over the title of best sedan in the world from Jaguar, what praise can be left for the Series III XJ Jaguar?

Within minutes of returning a 380SEL and driving off in a five-speed Series III we had the answer, one which left us once again filled with admiration for the refinement of the Jaguar's suspension and for the standard-setting ride and quietness that has existed since the first XJ sedan back in 1969 and is still unequalled by any other car, even the mighty Mercedes.

No other car combines suppleness of its ride, nor the quiet hushed way in which the tyres pass over potholes or joints in the road, with such marvellously controlled body movements to make driving and riding in the Jaguar journeys of such smoothness that it is impossible not to wonder how Jaguar's engineers, committed to massive weight reductions in the development of the new car due in 1984, can ever improve on these aspects of their car's dynamics.

So the Series III Jaguar still deserves special consideration for its charm and its abilities even if it is, inevitably, beginning to reflect its age in many obvious ways.

The series III was released in Britain in early 1979 and arrived here later the same year. Superficially it was a simple facelift but few people realise that the only body panels carried over were the floorpan, boot and bonnet. The roofline was raised, the windscreen made more steeply raked, the side windows made deeper and the front quarter vent windows were deleted while black wraparound bumper bars, recessed door handles and a new grille with vertical bars featured. Mechanically, the most important change was the adoption of electronic fuel injection for the classic twin overhead camshaft six. As well as boosting power the Lucas/Bosch injection ensured that starting was easier — although you must remember not to touch the accelerator — and made the engine smoother and quieter.

Power went from 134 kW at 4500 rpm to 152.9 kW at 5000 rpm while torque remained the same at 314 Nm, but at 1500 rpm instead of 3000 rpm, so the torque curve has been flattened out considerably. The increased power is reflected in the performance figures, for the new car is substantially quicker than the Series II 4.2 Jaguar we tested in March, 1979.

For this road test we drove both the automatic and the manual versions of the Series Three, the manual having become available in limited numbers late last year. The five-speed gearbox comes from the Rover 3500 and is a no cost alternative to the Borg Warner type 65 three-speed automatic. The recent return to three-pedal versions of expensive sedans is an interesting sidelight to the luxury car market. Until recently only BMW has been offering change yourself gearboxes but now we have Rover, Jaguar, Volvo (with the 264) and even Mercedes-Benz (with the 240D) selling cars that require the driver to exercise his left foot and arm.

Of course they only represent a tiny percentage of sales and perhaps exist only to improve the fuel consumption figures but they are a welcome addition for the variety they offer. But if our experience with the Jaguar is typical then it is only going to be the hard core of enthusiasts who will be happy with a clutch and gearbox for the very simple reason that they are somehow out of character with the rest of the car.

Jaguar's XJ is so gentle, so soft and unruffled that the gearbox, which works well in the Rover, comes across as a notchy, slightly heavy shift in the Jaguar. Worse is the clutch which is definitely heavy and bites home suddenly so that gear changes require considerable concentration if a sharp, jerky shift is to be avoided, while the clutch must be slipped to get the heavy car away from standstill smoothly. Certainly on the open road it is very pleasant to wind the big twin cam six out to the redline in third and pull the lever through to fourth before pushing up and away to fifth. But in other conditions the weight of the two key controls makes it all seem unnecessarily difficult when, with only a small drop in performance, it could all be done for you.

Perhaps this seems to go against the grain of everything this magazine stands for and we had high expectations the five-speed Jaguar sedan would be a highly desirable machine, but every staffer who drove the car, including the younger enthusiasts, complained about the physical and mental effort involved in driving the car.

So, despite the nasty automatic selector which all too easily allows the driver to push through from D to N but won't allow an easy shift back from D to 2, it is the Borg Warner box that gets our vote. Left to its own devices it changes up at 4300 rpm and 5000, which is the redline, and it is only marginally quicker to run the engine out to the red in low. The performance figures tell the story. The old Series II Jaguar covered the standing 400 metres in 18.4 seconds while the Series III automatic takes 17.8 seconds (in Held, in Drive range it is only 0.1 second slower) and the manual version 17.5 seconds. And if you check our acceleration in the gears times then you will discover that in some of the lower ranges the automatic in kickdown is actually faster than the manual; even in third gear.

The five-speeder is quicker off the line in ultimate terms but since this requires such perfect co-ordination we believe the automatic has more useable performance in the hands of most drivers.

Perhaps the only real advantage of the manual is that its economy potential is

greater. The manual uses a 3.31 final drive ratio compared to the automatic's 3.07 but with an overdrive fifth gear its top gear km/h per 1000 rpm is 44.2, while the auto's is 39.7, so that at 110 km/h the manual is pulling 2500 rpm and the automatic 2800 rpm. There is a definite benefit with the manual if you drive for maximum economy but in general motoring the margin is going to be inconsequential.

The fuel injection has certainly given the now almost 35 year old engine a new lease of life. Of course, it is a relatively slow revver but the harsh note of the old carburettor engine above 4500 rpm has been virtually eliminated so that now it never feels stressed although it doesn't have the ultimate smoothness of the Jaguar V12 or a Mercedes-Benz six or eight. Once more it is the automatic which feels calmer, the manual gear lever developing a slight buzz above 4500 rpm that intrudes into the relaxed atmosphere.

One other aspect of the manual which counts against the car is the cramped pedal area. There is very little room between the three pedals and it is possible to catch your right shoe under the brake pedal when lifting off the accelerator.

XJ's suspension remains unrivalled for its combination of smoothness and quietness at low speed and wonderful high-speed control (above). Cabin (left) is low, snug and very inviting.

Otherwise, the Jaguar is much as before. The steering, despite changes to the valving to reduce the amount of assistance, is still too light although we feel there might be slightly more feel. The steering feels lower geared than the 3.4 turns lock to lock would indicate, especially when turning into a corner when the car gives the impression of developing excessive initial understeer. Of course it doesn't, and once lock is applied the car goes where it is pointed, bumps seemingly having no effect. The roadholding is excellent and the ride unsurpassed. Interestingly, the automatic was fitted with Dunlop SP Sport tyres while the manual was on Pirelli P5s. The Dunlops are definitely quieter but the handling is improved with the P5s.

The brakes remain outstanding with a progressive pedal and strong resistance to brake lock-up but in a crash stop it will lock a wheel, something that's not possible on the ABS equipped S-class Benz, a feature that is to be incorporated into the new Jaguar due in 1984.

If there are more pleasant cars to ride in than the Jaguar then they have been kept far from us. For four adults the cockpit is roomy and comfortable as long as the driver doesn't mind sitting low and the rear seat occupants don't try to wear hats. The high waistline and scuttle are

SPECIFICATIONS

MAKE	**JAGUAR**
MODEL	**Series Three 4.2**
BODY TYPE	**Four door sedan**
COLOUR	**Red**
PRICE: Basic	**$45,380**
ENGINE:	
Cylinders	Six, in-line
Valves	Twin, ohc
Carburettor	Lucas electronic fuel injection
Compression ratio	8.7:1
Bore × stroke	92.07 × 106 mm
Capacity	4.235 litres
Max Power	152.9 kw at 5000 rpm
Max Torque	314 Nm at 1500 rpm
TRANSMISSION:	
Type	Three-speed, fully automatic, (five-speed, manual)

Ratios: (Manual figures in brackets)

	Gearbox	Overall	km/h-1000 rpm
First	2.40:1 (3.321:1)	7.37:1 (10.992:1)	16.6 (11.1)
Second	1.46:1 (2.087:1)	4.46:1 (6.908:1)	27.2 (17.6)
Third	1.00:1 (1.396:1)	3.07:1 (4.621:1)	39.7 (26.4)
Fourth	(1.000:1)	(3.31:1)	(36.8)
Fifth	(0.833:1)	(2.757:1)	(44.2)
Final drive	3.07:1 (3.31:1)		

CHASSIS:	Unitary
SUSPENSION:	
Front	Independent, lower wishbones, coil springs, anti-roll bar
Rear	Independent, lower wishbones, coil springs
STEERING:	
Type	Power assisted rack and pinion
Turning circle	12.2 m
Turns lock to lock	3.4
BRAKES:	
Type	Disc/disc
DIMENSIONS:	
Wheelbase	2865 mm
Track, front	1480 mm
Track, rear	1495 mm
Length	4959 mm
Width	1770 mm
Height	1377 mm
Kerb mass (weight)	1830 kg
Ground clearance	178 mm
FUEL TANK:	91 litres
TYRES:	Dunlop SP Sport ER70VR15 (auto) Pirelli P5 205/70VR15 (manual)

two factors which announce the age of the body but we're surprised that more effort hasn't gone into cleaning up the interior. Nevertheless, it has an old English character many buyers will enjoy.

Automatic air-conditioning is standard equipment and so, too, is a top quality sound system. Leather and cloth upholstery is offered, electric windows are standard and so is central locking but the steering column stalks don't move back with the wheel when it is adjusted for reach and a height adjustment is not too much to expect in such an expensive car.

Changes to the Series III Jaguar have

PERFORMANCE

TEST CONDITIONS:

Weather	Cool, dry
Location	Castlereagh Drag Strip
Load	Two people
Fuel	Premium

SPEEDOMETER ERROR:

Indicated km/h	Actual km/h
50	50
70	70
90	90
110	109
130	128

FUEL CONSUMPTION ON TEST:

Check one (auto)	6.5 km/l (18.4 mpg) over 456 km
Check two (auto)	7.3 km/l (20.5 mpg) over 585 km
Check three (auto)	5.4 km/l (15.2 mpg) over 379 km
Average	5.7 km/l (16.0 mpg) over 2810 km
Check one (man.)	5.4 km/l (15.3 mpg) over 335 km
Check two (man.)	6.5 km/l (18.3 mpg) over 460 km
Average (man.)	5.9 km/l (16.8 mpg) over 990 km

MAXIMUM SPEEDS IN GEARS:

Drive

First	75 km/h (4500 rpm)
Second	136 km/h (5000 rpm)
Third	190 km/h (4800 rpm)

Held
(Manual figures in brackets)

First	83 (55) km/h 5000 rpm
Second	136 (88) km/h 5000 rpm
Third	190 (132) km/h 5000 rpm
Fourth	(184) km/h 5000 rpm
Fifth	(195) km/h 4400 rpm

ACCELERATION Through the gears:

km/h	Auto Drive secs	Auto Held secs	Manual secs
0-50	3.9	3.9	3.6
0-60	5.1	5.0	4.6
0-70	6.3	6.0	5.8
0-80	7.9	7.6	7.1
0-90	9.5	9.0	8.5
0-100	11.3	10.9	10.3
0-110	13.5	12.9	12.3
0-120	16.0	15.6	14.6
0-130	19.1	18.7	16.9

In the gears:

km/h	Auto (kick-down) secs	Manual Second secs	Manual Third secs	Manual Fourth secs	Manual Fifth secs
30-60	3.4	3.4	4.8	7.9	12.0
40-70	3.3	3.4	4.8	7.2	10.6
50-80	3.4	3.7	4.8	6.7	9.9
60-90	4.3	—	5.1	7.0	10.0
70-100	5.0	—	5.2	7.6	10.6
80-110	6.0	—	5.6	8.2	11.4
90-120	6.7	—	6.0	8.6	12.4
100-130	7.4	—	6.8	9.2	13.4

STANDING START (0-400 m)
17.8 secs (auto) 17.5 secs (man.)

kept the car competitive and it remains a superb vehicle. If Jaguar can retain its many virtues in a much lighter and more modern body with new generation engines then life at the top for Mercedes is going to be a struggle. □

BRIEF TEST: Jaguar XJ Automatic

Time Bandit

By Peter Young

Jaguar sales are taking off world-wide. After 2500 km in an automatic transmission Series III it's not hard to see why.......

JAGUARS ARE MADE FOR wide open spaces, right? With this in mind Managing Editor Brian Woodward and I pointed the XJ6 Series III in the direction of Melbourne one cold July afternoon and let the good times roll.

We emerged from the leather and wood interior 900 kms and 10 hours later feeling tired, but happy. Well, more like glad actually. Glad that in this day of econoboxes and metric motoring mediocrity someone still makes a car capable of carrying four people and all their luggage from A to Z swiftly, silently and safely. It was an incident-free trip, down the much-improved but still at times lethal Hume Highway. A demanding schedule in Melbourne delayed our departure until the afternoon two days later. After shrugging aside the calm of Melbourne traffic we were headed North again, tanks full and the road running clear and straight in front. The Jaguar gobbled up the distance, covering 455 km in one 4 hr and 20 minute stretch without pausing or doing anything to attract the constabulary. At these speeds it returned an average of 14.5 litres per 100 km (19.4 mpg). Good, but not in the realm of BMW's high-speed fuel efficiency. We arrived back in Sydney after 2500 kms still delighted with the big Jag's ability to conquer time and distance.

Our test car was the automatic version of the Series III featured in MOTOR April. The main difference between the two was the automatic transmission which provided a weight saving of 70 kg (154 lbs). At 1760 kg kerb weight the Jaguar is still not a light car and with only 152.9 kW emerging from its fuel-injected straight six, performance in the lower speed ranges is distinctly leisurely. Part of the problem is of course the engine itself. It was winning races when I was still in nappies and it's long-stroke design and timid compression ratio of 8.7 to 1 don't exactly make for a modern powerhouse. However, the handy 314 Nm of torque helps, until it's fed into the leisurely Borg-Warner 66 automatic transmission that is. This auto box has all the worst traits of early English autos. It jumps into second gear almost as soon as you are moving, changes down reluctantly and has to be positively browbeaten into changing down on throttle kickdown. In a word, unresponsive and not befitting the Jaguar's other highly responsive controls and character.

Once moving at touring speeds, however, the Jaguar is deceptively fast. The driver must keep a constant watch on the speedometer as it is all too easy to creep up into highly illegal velocities. This is a car which makes a mockery of our open-road speed limits. This was brought home to us on the Melbourne trip. Just before Yass a semi-trailer shed one of its retread casings about a kilometre in front of us. After stopping to remove the bulky tread from the road, we were able to catch the semi-trailer within 15 km and 10 minutes. In the process the Jag's speedo reached over 160 km/h repeatedly. At these speeds the suspension seemed to function even better than at lower speeds and the benefits of the car's tremendously high gearing became apparent. This is a car built to move constantly at velocities rarely achievable even in wide-open Australia. This Jaguar isn't perfect, however, despite what you may have read or not read in other magazines. The steering (power-assisted rack and pinion) is still highly overboosted and insensitive. The handling is generally impeccable, with gentle understeer that is difficult to gauge at first because of body roll and the insensitive steering. However, at the limit the Jaguar shows its claws by moving into roll oversteer if the throttle is snapped off smartly in the middle of a fast bend. This shows itself with the rear suspension "jacking up" slightly and transmitting a distinctly insecure feeling to the driver. The speeds at which this occurs are very high however, and would rarely be achieved in everyday driving. The brakes need a hefty shove to create maximum retardation, but are well-modulated and sensitive in comparison with the feel available through the steering wheel.

Suspension in the XJ Series III is an example which most other makers could well use as a starting point for conventional (steel spring) suspension systems. Unlike nearly all Japanese suspension, the Jag has adequate travel and excellent control of the travel that exists. Massive coil springs at four corners with twin shock absorbers on the rear corners ensure precise control under nearly all conditions. I say *nearly* because despite what I have read and you have read in other magazines, the Jaguar's ride is far from perfect and undeserving of the many accolades showered upon it. The main criticism is in the area of suppression of bump-thump and judder from catseyes, joining strips and small, sharp-edged potholes. Over these the Jaguar shows itself to be no better (or worse) than many other European cars.

Many other aspects of the Jaguar are beyond criticism. The comfortable leather seats with adjustable lumbar control can be adjusted to suit most. The automatic "set and forget" climate control could probably only be bettered by some American luxury cars and Rolls-Royce. The instrumentation is comprehensive, although the highly polished wooden dashboard is not to everyone's taste these days. The Lucas halogen headlights deserve special mention. On full beam they will illuminate road signs about 1.6 km in front of the car. If they were any more powerful on high beam they would probably burn the leaves off the trees. Also, cutoff on low beam is not as severe as on many other European cars and still provides plenty of illumination. Other good interior touches include the twin electrically-adjustable exterior mirrors and the general level of fit and finish.

Nothing is ever done by halves on a Jag. On reflection, Jaguar seems to do everything in pairs — twin fuel tanks, twin exhausts, twin front and rear fog lamps, twin headlights (four altogether), twin mirrors, twin main instruments and twin Jaguar badges on the flanks. Perhaps because of this balance the XJ's styling still draws envious glances wherever it goes

That is until you pull into a petrol station and the owner comes out rubbing his hands at the thought of the 90-odd litres needed to slake the Jag's thirst. There's something distinctly disturbing about watching *two* petrol pumps going hell for leather to fill up just one car. The XJ6 as we know it now is due to be replaced in the near future with another model, to be known as the XJ 40. The existing car is beginning to show its age (it was first introduced in 1968) in some areas. However, there's no denying the basics of good design in the XJ — most of the weight is within the wheelbase and most of that weight is also down as low as the designers could get it. Both are contributing factors to its longevity in the face of competition from more recent designs.

Since the recent Jaguar price reductions in Australia, the XJ now stands well below the top Benzes and BMWs in the price stakes. With improved quality and reliability, plus the traditional Jaguar attributes of silence, safety and speed, that should be enough to keep it selling well until the new model arrives. The quality improvement should also help stabilise the secondhand value of Jags, where they used to suffer alarming depreciation. □

RAPPING

A JAGUAR DOESN'T travel the roads like an ordinary car. It doesn't wind its way through the streets of Melbourne to press-on up the Hume-fight of quick-slow-quick-quick-slow. You enter the Jaguar through your $40,000 thick wallet (or through your job if you're a pauper living like a prince in this line of work). The car hovers slightly above the ground as the engine distantly grumbles into life. Like the downstairs butler who has been asked to prepare a hot cup of Horlicks long after bedtime. Then the world turns beneath the Jaguar.

Towns flit by. That fauna which can survive the Hume's noxious gases flash past with complete disdain for Harry Butler. Finally the Jaguar sets its occupants down in another time and another place. Surely it isn't perfect? It isn't. It's just better than anything else. Rich, plush, comfortable, swift, competent, arrogantly inconsiderate of our dwindling fuel resources and above all, luxurious. Such decadence should be banned. Or compulsory. — Brian Woodward.

Jaguar XJ 4.2 Automatic Series III

ENGINE

Location	Front
Cylinders	Six in-line
Bore X Stroke	92.1 x 106 mm
Capacity	4235 cm^3
Carburetion	Lucas Electronic fuel injection
Compression Ratio	8.7 to 1
Fuel Pump	Electric
Valve Gear	Double OHC
Claimed Power	152.9 kW at 5000 rpm
Claimed Torque	314 Nm at 1500 rpm
Maximum Recommended Engine Speed	5000 rpm
Specific Power Output	36.1 kW/litre

TRANSMISSION

Type	Three-speed, automatic (Borg-Warner 66)
Driving Wheels	Rear

Gearbox Ratios

Gear	Ratio	km/h-1000 rpm	Max Speed	At (rpm)
First	2.4	16.5	88	(5000 rpm)
Second	1.45	27.33	136	(5000 rpm)
Third	1.00	39.6	198	(5000 rpm)

Final-Drive Ratio	3.07 to 1

SUSPENSION

Front	Independent by upper and lower semi-trailing control arms and anti-roll bar
Rear	Independent by lower control arms, with drive shafts acting as upper links with twin coil springs
Wheels	Dunlop Steel 6JK x 15
Tyres	Pirelli 205/70 VR 15

BRAKES

Front	284 mm discs
Rear	263 mm discs

STEERING

Type	Power-assisted rack and pinion
Turns, lock to lock	3.3
Ratio	17.6 to 1
Turning Circle	12.2 metres

DIMENSIONS AND WEIGHT

Wheelbase	2865 mm
Front Track	1480 mm
Rear Track	1495 mm
Overall Length	4959 mm
Overall Width	1770 mm
Overall Height	1377 mm
Ground Clearance	178 mm
Kerb Weight	1760 kg
Weight to Power	11.51 kg/kW

CAPACITIES AND EQUIPMENT

Fuel Tank	Two x 45.5 litres
Cooling System	18 litres
Engine Sump	6.8 litres
Battery	12V 66AH
Alternator	75 Amps

FUEL CONSUMPTION

Average for Test	14.5 litres/100 km
Best Recorded	12.9 litres/100 km
Worst Recorded	16.2 litres/100 km

ACCELERATION

	Held	Auto
0-60 km/h	5.48	5.44 secs
0-80 km/h	8.02	7.93 secs
0-100 km/h	11.34	11.80 secs
0-110 km/h	13.46	13.85 secs
0-120 km/h	15.88	16.23 secs
0-130 km/h	18.21	19.01 secs
Standing 400 metres	126.85 km/h	18.01 secs

CHECKLIST

Alloy Wheels	No
Adjustable Steering	Yes
Air-conditioning	Yes
Carpets	Yes
Central door locking	Yes
Clock	Yes
Intermittent Wipers	Yes
Laminated Screen	Yes
Petrol-filler lock	(two) Yes
Power Steering	Yes
Power Windows	Yes
Radio	Yes
Tape Player	Yes
Rear-window Wiper	No
Remote outside mirror adjustment	Yes
Sun Roof	No
Tachometer	No

SPEEDO CORRECTIONS

Indicated	Actual
40	40
60	60
80	80
100	99
110	109

LIST PRICE	$40,950
As tested	$41,950

This includes metallic paint at $1000

You must be Jochen!

Gerard Sauer visits a bespoke Jaguar dealer with a little bit extra to offer

JOCHEN Arden, the owner of a prosperous Daimler Jaguar dealership in Kleve, Germany, is a man with a vision. He likes to think that current day Jaguars can compete on equal footing with Mercedes Benz on MB's home ground.

Now, you may have noticed the odd Mercedes both in Germany and over here that has a bit extra in the way of body kits and tuning parts fitted. But how often do you see a Jaguar similarly adorned? Arden thinks this is all wrong and has actually done something about it. But first, more about his dealership.

Having only recently become a Jaguar dealer (at the beginning of 1982) has meant that he has none of the bad traits of the old guard, those who were brought up to think that having a Jaguar is enough of a privilege for an owner to endure any amount of aggravation. As Jochen put it; "The hardest job we have got at present, is to convince the Mercedes owner that the Jaguar is now a reliable car." To get this message across Arden set about the task with vengeance. First of all, good premises were found with ample showroom space, and a liberal helping of workshop area and stores. Then (much like a Mercedes dealership) through paying good wages he attracted first class mechanics. These were then further trained in the art of maintaining a Jaguar. To back up the generous workshop facilities, and to dispel any fears of cars sitting in the workshop for weeks on end waiting for parts, he built a stores and, twice weekly, sends his storeman over to the importers about 100 miles away to guarantee adequate parts supplies.

Says Jochen "to break down the old image of an exclusive but rather troublesome car, you have to take some thorough measures."

It is perhaps significant that the last three dealers between them, over a period of five years, managed to sell just five motorcars. In the first ten months of becoming a Jaguar dealer Jochen Arden sold seven in his area. And this year he hopes to make that 20. But, while he recognised the shortcomings of the old network, he also realised that, in this day and age, there are a number of people for whom a 12 cylinder 5.3-litre limousine is not exclusive enough.

Arden modifications

When we heard about Arden's work on Jaguars, we arranged to drive over and try them for ourselves.

The first thing that struck us was how unusual it was to see a Jaguar looking as good as this — just a few subtle mods made the difference between 'just another Jaguar' and a head turner. So what exactly did they do?

Taking a standard Jaguar XJ6 or V12 (the name Daimler is no longer used in Germany, as it is considered too confusing when you already have Daimler Benz on the doorstep) the Arden technicians start by lowering the car approximately 3cms which is achieved with the help of a new set of springs. This measure goes together with the fitting of a set of Koni dampers, specially made for the Arden Jaguars, and the valves adjusted to a sports setting. Then the standard wheels are exchanged for 16" light alloy rims of 7J width, and these are shod with 225x50x16 Pirelli P7 tyres. Finally some rather subtle side skirts are fitted together with a nice looking front spoiler.

This little lot, combined with the lowering of the car, adds a subtle touch of aggression. But as you can see from the photographs it is not over stated or tasteless. Incidentally, the Arden people consider that a rear-spoiler would be an aesthetic eye-sore. A view I can wholly agree with. This is further made unnecessary because, even at the highest speeds, the car is still a very stable performer.

To add to these measures, it is further possible to have all the chrome work on the car treated with an expensive looking matt black as with the white car in the photographs but, as this only works on certain colours, this is left up to the customer. To go with the new sporting image is an Arden exhaust system made from stainless steel which is modified from the centre backwards. This gives the engine a more sporting tone. It also allows, on the V12 engine, the fitting of dual exhaust pipes per side at the rear.

By relieving some of the restriction in the standard exhaust system it adds 15bhp at the rear wheels. And so typically for a V12XJ saloon, the 0-60 time is lowered from a standard 8.8 sec to 7.6 and the 0-100mph from 20 sec on the standard model to 17.2 on the Arden exclusive Jaguar. All these modifications can of course also be applied to the XJS.

As you can see from the photographs this car too, benefits greatly from the subtle touches available from the Arden range.

Interior touches

The interior has come in for some subtle re-touching including optional installation of the most expensive Recaro seats you are ever likely to come across. Electrically adjustable in practically every mode as, an extra option, these seats can be fitted with an electrically heated cushion. On the XJS we tried these Recaro 'C' seats can also be equipped with a system which allows electric-pneumatic adjustment of the seating cushions to shape the seat to your own profile and comfort! Add to that the optional leather steering wheel of a smaller diameter and the Connolly leather roof lining, and there is something there for the most expensive of tastes.

Talking about expensive, you might have been wondering what all this costs. The standard car in Germany delivered to your door would cost £17,682.84 for a Jaguar 66HE. The front spoiler fitted and painted costs £495.34. The side skirts £464.38, the matt black chrome £913.28. Alloy rims cost around £600.60 and painting the rims colour-matched to the car costs £111.45. The Pirelli P7s are £743, Koni shock absorbers fitted £402.46, shortened springs fitted £464.38. Then the stainless steel exhaust, fitted, is another £495.30 and those Recaro C seats in the front, for two, fitted, £2,383.35. An optional Blaupunkt Berlin radio fitted is £773.97, making the total price of this package £25,530.35. Exclusive indeed.

XJ 66HE with the subtle work of Jochen Arden. Opposite, the matt black treatment looks best on the lighter colours

How they drove

At first it is difficult what to think of these modifications, because it is hard to fault the Jaguar's road behaviour, and equally difficult to imagine how to improve it. But all that is put into perspective, once you get on the road with the car. The XJS we drove (the property of Herr Baltz of the Weingarten Sports shop in Dusseldorf) behaved in an exemplary way. The wide tyres are obviously suited to this weight of vehicle, as they rather add to the feel, and the mild amount of bump and thump that comes through on slow driving is not in any way intrusive but rather tends to add to the feeling of contact with the road that the car creates.

The traditional lightness of the steering wheel servo system has also been somewhat damped and now feels much more realistic and in tune with what is happening on the road. Altogether a more solid feel which the sporting driver would appreciate, as the softness of the standard car tends to erode confidence rather than enhance it.

Although the cobblestones produced a definite rumble it was mute enough not to matter, and once at speed the legendary Jaguar quiet prevailed. The only criticism of the car I have is that the altered exhaust system, though producing a fabulous sound, also has an annoying boom period around 50mph. And this somewhat detracts from the car's refinement.

On the outside, the lowering of the car, together with the front spoiler and side skirts, has resulted in a slightly lower drag factor which, together with the 15bhp increase at the rear wheels, has meant a raising of the top speed by 6mph. I found that the car was stable and gave you that sure-footed feeling usually reserved for S class Mercedes.

Currently, Jochen Arden's men are working on a tuned version of the V12 engine. Their aim is to extract another 55bhp from the engine, and they reckon that is well within its capacity without in any way compromising its smooth running and reliability. Judging by the results so far, I think they will succeed. And that really begged the last question. Did he think you could better a Jaguar? "It is not a matter of improving the car, but more a case of giving the customer the opportunity to adapt the car to his liking, to widen its appeal and to add to the exclusiveness and prestige of the make."

Judging by the number of heads that turned when we drove the car through a busy Dusseldorf, I think he's right.

Jaguar XJ6 4.2 5-speed

£15,997 ☐ *Max: 131 mph* ☐ *0-60 mph: 8.6 sec* ☐ *18.3 mpg overall*

EVER since 1978, when Jaguar first catalogued the 4.2-litre XJ6 with a five-speed manual gearbox option using the 77mm Rover SD1 transmission, as opposed to their own four-speed, we were keen to test the car. Well over 80 per cent of all XJ6s sold are automatics. It was also found that the Rover unit, whilst adequate for its original job, had strength and refinement deficiencies in its new Jaguar role. Nevertheless, as enthusiastic supporters of the use of five-speed overdrive type (not four-speed plus electric overdrive, as first used on the XJ), we continued to press Jaguar for a test car. Following further development, it eventually arrived last year; we tested it – and also used it successfully to re-enact something of the first Land's End to John o' Groats in top gear run (see page 15) – then were forced by pressure on space to hold over the test until now.

As supplied, the test car arrived with the 3.31 final drive; there are also 3.07 and 3.54 options available. This gears it at 23.15 mph per 1,000 rpm in the direct fourth, and 29.23 in the overdrive fifth. The last automatic XJ6 we tested (29 December 1979) had the 3.07 final drive. However, the combination of its epicyclic gearing and maximum torque converter multiplication (2 to 1) allows its

MANUAL GEARBOX overdrive five-speed version of Jaguar's 4.2-litre XK-six engined saloon, more often seen in three-speed automatic form, last tested 29 December 1979. Since these test figures were taken, this model has better stereo radio/cassette unit – otherwise mechanically identical to current model.

MODEL TESTED: XJ6 4.2 five-speed, four-door, 4,235 c.c. 6-cyl 2 ohc engine 205 bhp (PS-DIN).

FOR:
- *Excellent performance*
- *Superb cruise refinement*
- *Impeccable ride/roadholding*
- *Good value*

AGAINST:
- *Engine comparatively fussy at top end*
- *Clutch heavy*
- *Poor pedal and handbrake arrangements*

maximum spread of gearing (from torque converter stall in low to converter near lock-up in top), which respectively amounts to 5.1 to 24.7 mph per 1,000) to comfortably embrace the equivalent manual gearing between first and fourth (6.97 to 23.15 mph per 1,000).

So it is interesting to compare the full range of standing start acceleration figures for the two cars, theoretically with equal power – 205 bhp (PS-DIN) – and equal weight. As the table shows, the manual car's ability to use higher revs for the start, of which more later, pays off handsomely up to 40 mph, from there the automatic does unexpectedly well to keep within the same 1.1 to 1.4 sec gap up to 80 mph, presumably due to the five-speed's extra gearchange, one of which is the slow second to third one. Then the manual's higher efficiency pays off increasingly, to such an extent that it was possible to measure acceleration to 120 mph within the mile-long MIRA twin horizontal straights with the manual but not the automatic.

Standing start acceleration

	1979 automatic	**1982 Manual**
mph	sec	sec
0-30	3.6	2.8
40	5.6	4.2
50	7.4	6.3
60	10.0	8.6
70	13.3	12.1
80	16.9	15.6
90	21.9	20.2
100	29.4	26.5
110	38.5	34.6
120	–	45.6

In fact, standing starts on the manual car are distinctly nerve-racking, because of the alarming juddering tramp set up from rear sub-frame to engine mounts by racing type wheel-spin. The clutch normally takes up very well unless subjected to this sort of abnormal abuse; its only bad point is an old one, familiar to owners of all previous manual Jaguars – excessive pedal effort. Replacing the undeniably good vibration-absorbent properties of a torque converter with a clutch also transmits the famous XK top end crankshaft vibration very much more faithfully than in the automatic.

Returning to the five-speed's performance, if one observes the lowly 5,000 rpm red line on the rev counter – which is also the peak power speed – when changing up, each gear's maximum speed corresponds to 35, 55, 83 and 116 mph, the revs dropping to 3,150, 3,300 and 3,600 rpm respectively. The low peak is more than made up for by the six's extraordinary flexibility, which masks both the rather high first and the low second. In spite of the high-sounding overdrive, it was possible to take acceleration figures in fifth from 10 mph – 340 rpm – and even, for our top gear run attempt, to start from rest in fifth, with the help of only a little clutch slip.

In practice, with the 3.31 final drive, the car is not allowed to make full use of its potential long-leggedness, as the near identical maximum speeds in fourth and fifth demonstrate. From previous Jaguar recommendation, we happily risked exceeding the red line in fourth until the maximum in that ratio had been reached, at a considerable 600 rpm – 12 per cent – beyond the peak, whilst fifth turns out to be only a 10 per cent overdrive. Cars of this character and power-to-weight ratio are easily capable of pulling much higher overdrives, giving around 20 per cent overgearing at maximum speed. The ideal in the Jaguar's case would be a higher final drive, plus a lower first to preserve maximum hill climb capabilities.

In manual form, car retains awkward-to-reach umbrella handbrake – a pity when it is regularly needed (it isn't on automatics) – but otherwise has excellent control layout, including correctly arranged horn switch, using centre of steering wheel spoke

Even as it is however, the five-speed XJ is a very much less thirsty machine than in automatic form. The worst interval consumption seen was 17.2 mpg – after correction for an excessively over-reading mileometer (6 per cent out) – whilst we twice saw between 23 and 25 mpg on long runs. The overall test consumption – which does not include the Land's End-John o' Groats run, but is as usual made a lot worse by the performance testing part – works out at 18.3 mpg, against the automatic's 16.8. This improves the car's range usefully, to well over 300 miles on the two 10 gallon tanks.

On the Jaguar's prized refinement, there is a small loss felt if one uses the top end a lot, when some irritating engine harshness spoils the car. Here, as with the heaviness of the clutch, the car certainly shows its age – something the AJ6 24-valve six should put right in the next-generation saloon. In the present car's defence however, owners will agree that you really do not need to venture beyond 4,500 rpm often, thanks to that lovely flexibility, which the automatic always wastes.

The rest of the car is as it always has been – very fine steering, superb roadholding in all circumstances, a pleasingly firm yet absorbent type of ride, spoiled slightly by a degreee of transverse rocking, exemplary stability and balance, and first class brakes.

Coming back to the manual quality of the design, the clutch weight is doubly objectionable because it "deharmonizes" the pedals, in its great contrast to the excellent accelerator and brake pedal loadings. The pedal positions are not ideal, with the accelerator considerably below the brake, yet close enough for one occasionally to catch the welt of one's shoe underneath the brake when decelerating. One notices too the awkwardness of the umbrella handbrake, since its use is essential where on the automatic it was not. Long-legged drivers all demand extra rearward adjustment for the seat; for up to 6ft occupants, space is good however. Vision is very good, as ever; sound, in the shape of the somewhat mean AM-only radio fitted to the test car, is not up to its surroundings.

The five-speed XJ6 does show signs of something rare in present Jaguars – insufficient development. It is however still wonderful to be able to enjoy the (mostly) great smoothness, refinement and flexibility of that lovely power unit. Several of *Autocar's* testers agree that, were they able to afford it, the manual XJ6 is a very desirable machine – and what a waste of the even greater sweetness of the V12 that one cannot have a high-geared manual XJ12.

Performance

MAXIMUM SPEEDS

Gear	mph	kph	rpm
Top (mean & best)	**131**	211	4,500
4th (mean & best)	**130**	209	5,600*
3rd	**83**	134	5,000
2nd	**55**	89	5,000
1st	**35**	56	5,000

*see text

ACCELERATION

FROM REST

True mph	Time (sec)	Speedo mph
30	**2.8**	33
40	**4.2**	45
50	**6.3**	56
60	**8.6**	66
70	**12.1**	77
80	**15.6**	89
90	**20.2**	101
100	**26.5**	112
110	**34.6**	123
120	**45.6**	133

Standing ¼-mile: 17.2sec, 84 mph
Standing km: 31.3sec, 106 mph

IN EACH GEAR

mph	Top	4th	3rd	2nd
10-30	10.3	6.8	4.7	3.0
20-40	9.8	6.6	4.7	3.1
30-50	9.8	7.0	5.2	3.4
40-60	9.9	7.4	5.1	3.8
50-70	10.8	7.7	5.2	–
60-80	12.4	8.2	6.2	–
70-90	14.5	8.8	–	–
80-100	15.4	10.3	–	–
90-110	21.6	13.6	–	–
100-120	–	16.6	–	–

JAGUAR XJ6 4.2 MANUAL

FUEL CONSUMPTION

Overall mpg:
18.3 (15.4 litres/100km) **4.03** mpl
Autocar formula: Hard 16.5 mpg
Driving and conditions: Average 20.1 mpg; Gentle 23.8 mpg

Grade of fuel: Premium, 4-star (97 RM)
Fuel tank: 20 Imp. galls (91 litres)
Mileage recorder reads: 6.0 per cent long

OIL CONSUMPTION

(SAE 10W/40) 3,200 miles/litre

WEIGHT

Kerb, 34.6cwt/3,871lb/1,756kg
(Distribution F/R, 54.5/45.5)
Test, 38.3cwt/4,291lb/1,946kg
Max. payload, 900lb/409kg

TEST CONDITIONS:

Wind: 5-12 mph
Temperature: 48deg C (9deg F)
Barometer: 29.7in. Hg (1,007 mbar)
Humidity: 75 per cent
Surface: dry asphalt and concrete
Test distance: 1,591 miles

Figures taken at 8,600 miles by our own staff. All Autocar test results are subject to world copyright and may not be reproduced in whole or part without the Editor's written permission.

SPECIFICATION

ENGINE

Longways front, rear-wheel drive. Head/block al.alloy./cast iron. 6 cylinders in line, dry liners, 7 main bearings. Water cooled, viscous fan.
Bore, 92.07mm (3.62in.), stroke 106.00mm (4.17in.), capacity 4,235 c.c. (258 cu. in.).
Valve gear 2 ohc, chain camshaft drive.
Compression ratio 8.7 to 1. Breakerless ignition, Lucas/Bosch L Jetronic injection.
Max power 205 bhp (PS-DIN) (150.5kW ISO) at 5,000 rpm Max torque 236 lb.ft. at 3,700 rpm.

TRANSMISSION

Five-speed manual. Single plate diaphragm spring clutch.

Gear	Ratio	mph/1,000 rpm
Top	0.792	29.23
4th	1.0	23.15
3rd	1.396	16.58
2nd	2.087	11.09
1st	3.321	6.97

Final drive gear: Hypoid bevel, ratio 3.31.

SUSPENSION

Front, independent, double wishbones, coil springs, telescopic dampers, anti-roll bar. Rear, independent, lower wishbones and fixed length drive shafts, twin coil springs and telescopic dampers.

STEERING

Rack and pinion, hydraulic power assistance. Steering wheel diameter 15¾in., 3.3 turns lock to lock.

BRAKES

Dual circuits, split front and rear. Front 11.2in. (284mm) dia discs. Rear 10.4in. (264mm) dia discs. Vacuum servo. Handbrake, centre lever acting on rear discs.

WHEELS

Pressed steel, 6in. rims. Dunlop, Michelin, Pirelli tyres (Pirelli P5 radial on test car), size 205/70VR 15in., pressures F28 R26 psi (normal driving).

DIMENSIONS

Wheelbase 112.8in. (2,865mm); track, front 58.3in. (1,481mm), rear 58.9in. (1,496mm). Overall length 195.3in. (4,951mm) width 69.7in. (1,770mm), height 54.1in. (1,374mm). Turning circle 39ft 9in. (12.1m). Boot capacity 19.1 cu. ft.

What it costs

PRICES

Basic	£12,840.00
Special Car Tax	£1,070.00
VAT	£2,086.50
Total (in GB)	**£15,996.50**
Licence	£80.00
Delivery charge (London)	£201.25
Number plates	£28.75
Total on the Road (exc. insurance)	**£16,306.50**

TOTAL AS TESTED ON THE ROAD **£16,306.50**

Insurance Group 8/OA

WARRANTY

12 months unlimited mileage

JAGUAR XJ6 VANDEN PLAS

Mercy! The quality is not strained.

PHOTO BY THOS L. BRYANT

JAGUAR OWNERS, AS the novelist said about another problem we'd all like to have, are different from you and me. Instead of shouldering the burdens of great wealth, folks who have Jaguars own the most rewarding, elegantly sporting cars that ever drove anybody over the brink.

Every owner has a story to tell, the sort of epic that's great fun when it's all over and didn't happen to you.

There's a man who brags that he drove coast to coast in his new XJ6, record time, no tickets, "and it only caught fire once."

There are owners who buy their Jags two at a time, to be sure one is available when the other is in the shop.

There are stories about cooked engines, doors that mysteriously locked the owners inside, things that went bump in the night.

Nor is this merely folklore. This magazine surveyed Jaguar owners in 1978, and reported one of the worst service records in survey history. One third of the XJ6s had cooling problems. Another third had electrical malfunctions. Another 30 percent suffered engine troubles.

AT A GLANCE	Jaguar XJ6 Vanden Plas	Audi 5000S Turbo	BMW 533i
Price, base/ as tested	$34,700 $34,700	$22,250 $22,250	$30,305 $30,305
Curb weight, lb	4250	3090	3160
Engine/drive	inline-6/rwd	inline-5/fwd	inline-6/rwd
Transmission	3-sp A	3-sp A	5-sp M
0–60 mph, sec	12.3	10.6	8.3
Standing ¼ mi, sec @ mph	18.9 @ 75.0	17.8 @ 79.0	16.4 @ 85.5
Stopping distance from 60 mph, ft	152	159	149
Interior noise at 50 mph, dBA	63	67	67
Lateral acceleration, g	0.761	0.762	0.780
Slalom speed, mph	56.7	59.7	58.0
Fuel economy, mpg	17.5	18.0	19.0

XJ6 Vanden Plas: Classic styling and abundant luxury, plus new-found reliability.
5000S Turbo: A spacious and slippery sedan that combines speed and efficiency (1-84).
533i: The clear-cut performance leader among luxury sedans (2-83).

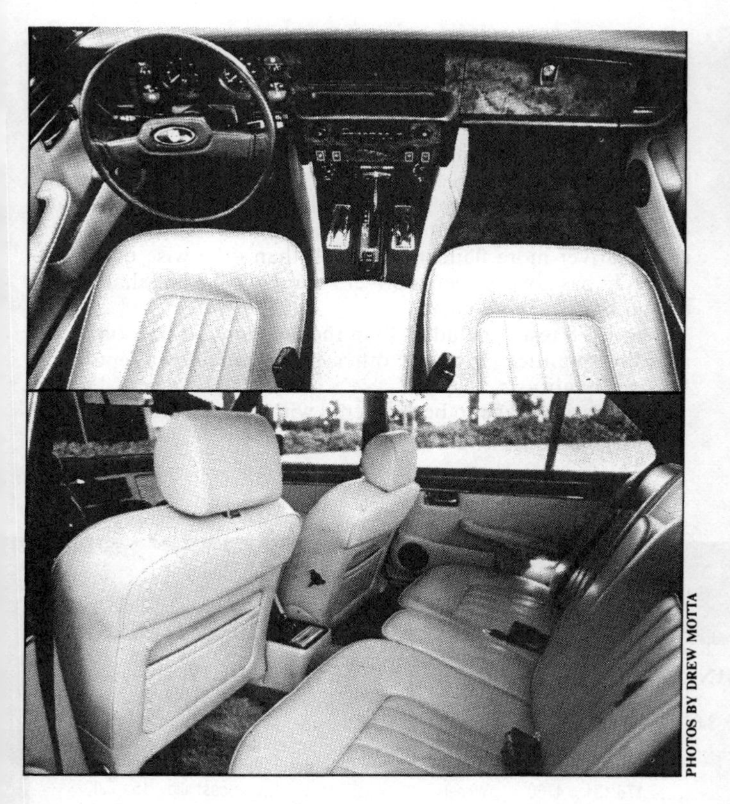

PHOTOS BY DREW MOTTA

Later that year we tested the then-current XJ6 Series III. The test involved a total of four examples of the model, just to have a running car long enough to get all the numbers.

Clearly, Jaguar had problems that couldn't be permitted. By no coincidence, the company became an independent factory again, after a period as a division within a conglomerate. Freed to manage, the new managers went through a series of new brooms, shaping up the quality control people and—probably the most important part—rejecting supplied parts unless and until those components met some tough standards.

We've got it whipped, Jaguar said, so would you like to try another XJ6? Skeptical but game, we said sure.

Presented here is the 1984 XJ6 Vanden Plas. The basic machine is a familiar package. The engine is the dohc inline-6 that established Jaguar as a sporting make back in 1949, as seen (in slightly different tune, of course) on the race courses of the world. It is now equipped with fuel injection, all modern electronics and backed by an excellent automatic transmission. The car itself is the standard 4-door sedan, now in its third basic body since the model was introduced in 1968. The XJ6 is the bread and butter, so to speak, of Jaguar, for easily appreciated reasons. It was a beautiful car when introduced, and it still catches the eye and warms the heart. It's dated only in that the rivals have lost some appeal in their trips through the wind tunnel, while the XJ still looks like a machine designed by people with style. It costs a lot and the money put into the car shows.

Vanden Plas refers to the optional interior, one of the few extras not included; air conditioning, AM/FM/cassette player, sunroof and trip computer are standard. The Vanden Plas consists of leather for all the interior panels rather than vinyl in some places, of the rear seat shaped so it's two defined seats instead of a semi-bench and of extra trim on the doors. And very nice it is, too.

Because of the previously mentioned reputation of the XJ, this report is divided into two parts.

First, proving the positives.

Jaguar's six is an incredible engine, evidence that if you do it right and keep improving, you don't have to come out with a new engine every year. In its 1984 form, it starts with a few spins, hot or cold, and simply runs. Smooth and quiet, with sufficient horsepower (176 bhp at 4750 rpm) and torque (219 lb-ft at 2500) to provide sporting performance—keeping in mind the car's 4440-lb test weight and luxury nature. We measured the Vanden Plas's 0–60 mph time at 12.3 seconds—measurably slower than the 10.6 sec we achieved with the XJ6 Series III in September 1980. Not surprising considering that the current luxury model is 310 lb heavier (test weight) and has a 2.88:1 final drive ratio compared with 3.07:1 for the earlier model.

Note here though that while the engine is a willing revver, happy to spin to the 5000-rpm redline, the transmission requires some driver attention. It's smooth and sure, but the final and internal gear ratios were chosen for easy cruising and good economy. The transmission shifts up at the first opportunity and automatic downshifts don't come as quickly or as firmly as the occasion can warrant. Use the lever, is the secret. First is perfectly safe for getting past that farm wagon; 2nd is a good passing gear for uphill sweeps around diesel sedans and barely mobile homes. Think Le Mans, in short, and the XJ will go along with you.

Handling displays the same heritage. It's surprisingly stable for such a luxurious ride. The driver gets plenty of input from the steering and—surprise—one can no longer get ahead of the power assist. At the limit there's predictable understeer. The car seems perfectly happy to flog around the skidpad (0.761g) or hustle through the slalom, which it does at a respectable pace (56.7 mph). The activity seems a bit undignified but driving at the limit reveals nothing worse than body lean. (As a positive owner note, one of that group has done rather more exploring of gravel and graded dirt roads than the Jag stereotype would predict. Good car for offroad, he says, plenty of ground clearance and the body structure is remarkably solid.)

The brakes plain work right. Respectable stops, little fade, no pulling or yaw. The response is linear; how hard you push controls how quickly you stop. No further comment needed.

The interior drew mostly raves. Mostly in that there was a consensus concerning the wood, which is perfectly in character. The test involved two cars, about which more later but it's not what you think. The actual test car shown had the Vanden Plas interior and there were some remarks about all that leather being a bit much. The fit and finish struck some as being too perfect, so good it looked like, well, like vinyl. The standard version drew no such remarks. Plain (relatively) or fancy, most occupants were right at home in the seats.

The interior has been reworked, for the better. The switches are where one expects them to be, and they feel right. Virtually the only debits on the interior chart were from an improperly installed remote control for one mirror and a shortage, literally, of space for the driver's right foot. The parking brake shaft runs just above and almost to the right of the accelerator pedal. Size nines and above hit this shaft. No big problem but one that the factory should have cured by now.

A surprise in several senses is the inclusion of a trip computer. Right, just as in cars from you-know-where. It's an unobtrusive system, no bells or whistles (although the relay was audible). At the push of several small buttons, it reports fuel used, mpg at that moment, mpg for the trip so far, miles covered, average speed, elapsed time and time of day. An away-from-home entertainment center, one could say. Several of the crew felt such a modern, space-age thing was out of character for the Jag. Others thought it was fun, glancing down while cruising the open road and being told you've gone 100 miles, at 20 mpg, averaging 62 mph, etc. If the video display palls, one can select time of day. A clock surely is permitted in a luxury sedan.

Now, part two, proving the negative.

Science says negatives can't be proved, that just because something hasn't happened doesn't mean it won't.

One can try, though. Jaguar's Draconian improvement program occurred between tests; that is, our most recent previous XJ6 test was back in the bad old days. In between we had a brief impression of the Vanden Plas option, but although it worked fine, it wasn't official.

Thus, the car shown here was subjected to extra scrutiny. The early notes in the logbook reflect past experience, along the lines of lovely car, too bad about those service problems, or I'd own one if it weren't for . . .

The first example in this test was driven hard for a month.

Nothing went wrong.

When that one went home, we swapped for another, an XJ6 with the standard interior. That one was used for another 1500 miles, as long and hard as good manners allow. It was kept idling in parking lots, crept through rush hour traffic, cruised across the desert at unprintable speeds.

Nothing went wrong. The needles stayed at the center of their dials. There were no mysterious noises, no showers of sparks, no puddles of steaming green coolant.

This is wonderful news. Later logbook notes reflect the pleasures of driving a car with character, one that somehow imparts to the driver more flair, more polish than otherwise displayed. The XJ6 has always been a driver's car. Now it's one that can be driven there . . . and back.

The 1978 test concluded, "On the one hand, it has some outstanding qualities . . . on the other, it is haunted by a reputation for poor quality.

"Our praises have to be tempered with our reservations."

Cancel our reservations. 

PRICE

List price, all POE	$34,700
Price as tested	$34,700

Price as tested includes std equip. (auto temp control, power adj driver's seat, AM/FM stereo/cassette, elect. sunroof, elect. window lifts, elect. mirrors, trip computer, central locking, cruise control, additional walnut interior trim and leather surfaces, floor rugs)

GENERAL

Curb weight, lb/kg	4250	1930
Test weight	4440	2016
Weight dist (with driver), f/r, %	54/46	
Wheelbase, in./mm	113	2870
Track, front/rear	58.3/58.9	1480/1496
Length	199.6	5070
Width	69.6	1768
Height	52.8	1341
Trunk space, cu ft/liters	14.3	405
Fuel capacity, U.S. gal./liters	23.6	89

ENGINE

Type	dohc inline-6
Bore x stroke, in./mm	3.63 x 4.17 .. 92.1 x 106.0
Displacement, cu in./cc	258 4235
Compression ratio	8.1:1
Bhp @ rpm, SAE net/kW	176/131 @ 4750
Torque @ rpm, lb-ft/Nm	219/297 @ 2500
Fuel injection	Lucas-Bosch
Fuel requirement,	unleaded, 91-oct

DRIVETRAIN

Transmission	automatic; torque converter with 3-sp planetary gearbox
Gear ratios: 3rd (1.00)	2.88:1
2nd (1.45)	4.18:1
1st (2.40)	6.91:1
1st (2.40 x 2.0)	13.82:1
Final drive ratio	2.88:1

CHASSIS & BODY

Layout	front engine/rear drive
Body/frame	unit steel
Brake system	11.1-in. (281-mm) vented discs front, 10.3-in. (262-mm) discs rear; vacuum assisted
Wheels	cast alloy, 15 x 6JK
Tires	Pirelli P5, 215/70VR-15
Steering type	rack & pinion, power assisted
Turns, lock-to-lock	2.8
Suspension, front/rear:	upper and lower A-arms, coil springs, tube shocks, anti-roll bar/halfshafts, lateral links, trailing arms, dual coil springs and tube shocks

CALCULATED DATA

Lb/bhp (test weight)	25.2
Mph/1000 rpm (3rd gear)	27.3
Engine revs/mi (60 mph)	2200
R&T steering index	1.12
Brake swept area, sq in./ton	193

ROAD TEST RESULTS

ACCELERATION

Time to distance, sec:	
0–100 ft	3.9
0–500 ft	10.3
0–1320 ft (¼ mi)	18.9
Speed at end of ¼ mi, mph	75.0
Time to speed, sec:	
0–30 mph	4.4
0–50 mph	9.1
0–60 mph	12.3
0–70 mph	16.5
0–80 mph	22.0

SPEEDS IN GEARS

3rd gear (5000 rpm)	117
2nd (5000)	88
1st (4500)	47

FUEL ECONOMY

Normal driving, mpg	17.5

BRAKES

Minimum stopping distances, ft:	
From 60 mph	152
From 80 mph	262
Control in panic stop	excellent
Pedal effort for 0.5g stop, lb	26
Fade: percent increase in pedal effort to maintain 0.5g deceleration in 6 stops from 60 mph	nil
Overall brake rating	excellent

HANDLING

Lateral accel, 100-ft radius, g	0.761
Speed thru 700-ft slalom, mph	56.7

INTERIOR NOISE

Constant 30 mph, dBA	58
50 mph	63
70 mph	69

ACCELERATION

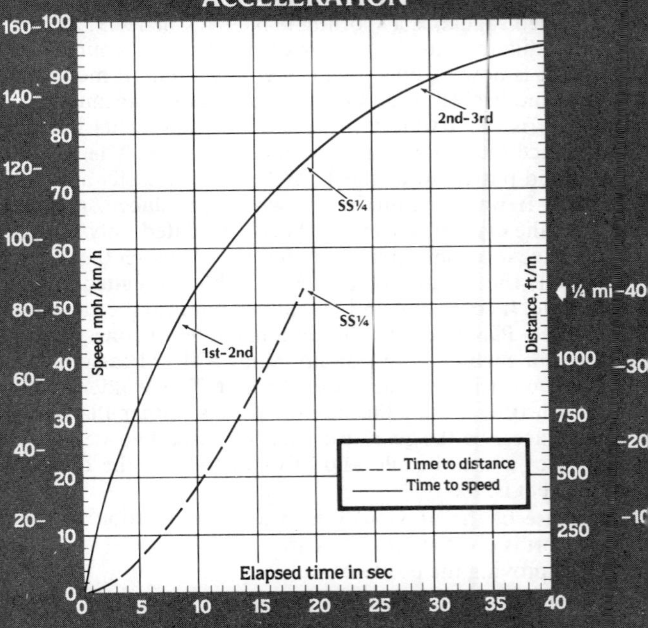

TEST MATCH

Audi 200 Turbo BMW 732i Jaguar Sovereign 4.2

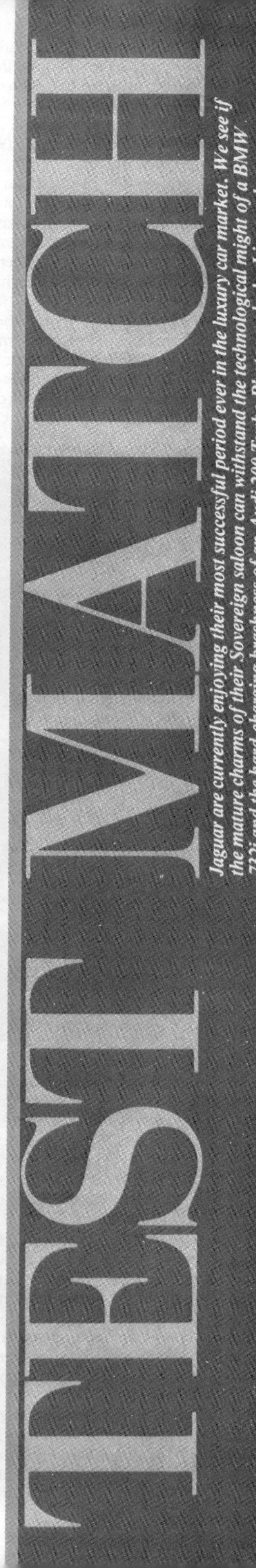

Jaguar are currently enjoying their most successful period ever in the luxury car market. We see if the mature charms of their Sovereign saloon can withstand the technological might of a BMW 732i and the hard-charging brashness of an Audi 200 Turbo Photographs by Lipscombe

FOR MOST true Brits, only one company sums up life in the fast lane – and that is Jaguar. That is why we rushed out in our thousands to buy shares in the newly privatised company; why the merest hint of a new Jaguar sports car or a return to Le Mans sets the blood stirring.

Yet until recently Jaguar survived – just about – more on its traditions and past glories than its current products, which were wonderful to drive but desperately unreliable to own. The transformation inside the company that has righted this sorry situation has been well documented, and Jaguar is now on the crest of a wave with booming export sales, especially in the USA.

That desperate '70s period allowed other companies to cash in on the lucrative luxury car market – one which has always stayed remarkably buoyant, despite fluctuating oil prices. German manufacturers, in particular, have come to dominate, with the autobahn expresses of BMW and Mercedes-Benz riding high.

Next year, Jaguar will have its all-new XJ40 saloons to challenge the Germans but for the moment it is competing surprisingly successfully with the elderly XJ6 and XJ12. Is its success, at home and abroad just down to a revival of patriotic spirit and the cheapness of the pound against the dollar, or can a 16-year-old car really compete with Germany's best?

To find out we have chosen the plush Jaguar Sovereign version of the six-cylinder saloon range and lined up BMW's 732i for comparison.

But Jaguar aren't alone in wanting to loosen the grip of BMW and Mercedes on this lucrative luxury market: Audi have been carrying out a carefully executed strategy, too, to move themselves into contention. And in contrast to the traditional status imagery of Jaguar and their German rivals, Audi have set out to woo the customer with an ultra-modern, high-tech approach, using the sleek aerodynamic body of their 100 model and Quattro-style turbocharging to create the very high performance 200 model. They have four-wheel-drive in mind, too, for the very near future.

So the Audi is the third – a challenger to the establishment and to the established standards of the other manufacturers, who will be going expensively wrong if the Audi formula proves itself a success.

The £17,013 Audi is the odd man out in every sense: the only one that is derived from a substantially cheaper model, the only one that is front-wheel-drive, the only turbo and so on.

Externally, it *is* the Audi 100: you must look carefully to spot the bigger lamps and narrower grille, slightly deeper spoiler, alloy wheels and the reflective strip across the rear boot panel. Aerodynamically, the flush-glassed body worsens from 0.30 to a 0.33 Cd because of the wider tyres and bigger spoiler, but it is still an impressive class leader – the BMW's poor 0.40 Cd shows the German company's relative disdain for aerodynamic sleekness, while the Jaguar's 0.44 was not bad in its day.

Mechanically, the 200's claim to stand apart from the cheaper model is its turbocharged version of the five-cylinder power unit they share. The 2,144cc engine is what Audi describe as a third generation turbocharging

development and it is substantially different from the old 200's turbomotor, with more power and torque.

The in-line engine now develops 182bhp at 5,700rpm, using a relatively high, 8.8:1 compression ratio and an intercooler behind the radiator. Programmed electronic ignition, incorporating a knock sensor, is used while the fuel injection is Bosch's K-Jetronic mechanical system.

Drive to the front wheels is through a five-speed manual transmission or the extra cost (£780) option of a three-speed automatic. We opted for the manual, but might have been tempted by automatic had the Audi offered a four-speed lock-up system that didn't sap so much of its performance potential.

Suspension is largely similar to the 100's but the use of bigger, 15in wheels with hefty 205/60VR15 Pirelli P6 tyres has necessitated re-tuning and both spring and damper rates are higher. Front suspension is by MacPherson struts and at the rear is a torsion-beam axle on coil springs with a rear anti-roll bar joining that at the front end for the 200.

Brakes see another big improvement, with ABS anti-locking standard on the 200 Turbo's ventilated front/solid rear disc set-up. Steering is powered rack and pinion.

What a contrast to all this the BMW makes. Its six-cylinder, overhead-cam engine produces slightly more power – 197bhp at 5,500rpm – by dint of the traditional means: a big capacity of 3,210cc. It gains in torque, too, but not a lot: 210lb ft at 4,300rpm compared with the Audi's 191lb ft at 3,600rpm. BMW use Bosch L-Jetronic electronic fuel injection and their Motronic management system is the electronic brains behind the engine.

There is a choice of transmissions for the 732i: five-speed manual or automatic four-speeder (£600 extra), while a third option, at a hefty £926, is the clever four-speed switchable automatic with Sport, Economy and manual change settings, that was fitted to our test car.

BMW have worked hard to turn their originally unhappy 7-Series cars into much better machines and the simple description of the suspension as being by MacPherson struts at the front and semi-trailing arms at the rear does not do justice to the amount of detail refinement carried out over the years. Brakes are all disc, ventilated at the front with ABS anti-locking a £972 option on the 732i but standard on the more expensive Special Equipment model. Alloy wheels (£541) on our car were fitted with Michelin XWV 205/70VR14 tyres.

Steering, unusually these days, is by recirculating ball and power assistance comes free. As you can see, it's difficult to put a firm price on the BMW: the list price of the 732i is £16,850 but what you pay in the end depends on what you want. We will go into specifications in more detail later but bringing the 732i up to a similar standard as the others here will cost well over £2,000, however you decide to spend it.

The Jaguar Sovereign stands at £18,995 and doesn't have the same lengthy table of mix and match extras as the BMW. Though the Sovereign as a model designation was new this year, the essential Jaguar formula remains little changed.

Powering the car is Jaguar's classic 4.2-litre twin-cam six cylinder engine that generates 205bhp at 5,000 revs and 236lb ft of torque at 3,750 revs. It is a modestly tuned, venerable engine whose 8.7:1 compression ratio is actually lower than the turbo Audi's. The only real concession to modernity is

Lucas/Bosch L-Jetronic injection, the adoption of which was one of a range of engine improvements when the Series Three was introduced in 1979.

That doesn't mean it isn't a fine engine; far from it, for the twin overhead camshafts and hemispherical combustion chambers are a model of classic design. But, by present standards it is rather a long stroke unit – with 96 × 106mm dimensions – and massively proportioned compared with the BMW six.

The Jaguar's transmission, as befits the traditionalist of the group, is a conventional three-speed automatic 'box, the Borg-Warner Type 66, though a five-speed manual is available at the same price.

Suspension is by the most sophisticated system of the three cars here. At the front there are double wishbones with coil springs and an anti-roll bar, while the independent rear end uses lower wishbones, twin coil springs and dampers each side, and fixed length driveshafts to form part of the suspension linkage. Brakes are solid discs all-round and the steering is by rack and pinion with power asistance. The Sovereign runs on distinctive drilled alloy wheels, fitted with 205/70VR15 Pirelli P5 tyres.

Despite their different specifications, the three contenders are surprisingly similar in size, from the Jaguar, longest at 195in to the Audi shortest at 189in. The Audi also has the shortest wheelbase, 106in compared with 110 for the BMW and 113 for the Jaguar.

However, kerb weights vary much more sharply: the Audi is the featherweight at 25.4cwt, the BMW not a lot more – 29.3cwt, but the Jaguar scales in at a massive 36cwt – middle-aged bulk that the new car will have to lose if it is to be competitive.

Top: rear passengers are well looked after in the Audi with good legroom, a comfortable seat and headphones for the stereo. Front seats lack proper height adjustment but are the most comfortable of the three here. Middle: BMW has typically firm seats, front and rear, and feels very spacious. Our car had electric adjustment for height (at front and rear of cushion), reach, recline and headrest height on the front seats plus a memory for storing three different positions. Bottom: Jaguar's cabin is a bit tight on headroom but is just as generous in other dimensions as the other two. The smell of leather is delicious but the seats aren't sufficiently well shaped to stop occupants slipping about

AUDI 200 TURBO

ALONGSIDE THE gleaming chrome of is rivals here, the Audi's smooth lines smack of understatement to the point of reticence. It is as sleek and good looking as that clean-lined body always has been, but Audi 100s are thick on the ground now – won't the 200 buyer want others to notice the extra £7,000-odd he's spent out?

The difference shows on the road, though: the modest performance of the 100 is left for dead by the scorching sprinting power of the turbo engine. The five cylinder starts and idles with that familiar irregular thrumming beat, then it warms to its task with a gentle run through the lower rev ranges, where it never feels flat or slow as some turbos can – just mild compared with what is to come as the revs spin past 3,000.

By then, the turbocharger is in full cry – there's no boost gauge on the minimally instrumented facia so you won't know exactly when full boost is reached – but from around half revs it really starts to fly. No blood-curdling wail or classic howl marks its progress up through the revs, just that distinctive five-pot roar, getting harsher as the 6,500 red line closes.

Forgive the lack of exciting sounds, for progress is exciting enough: the 200 can crack 60mph in 8.3sec and 100mph in 21.5sec on is way to a maximum speed of 141mph – a figure that scores the Turbo high marks in any saloon company and is far in front of the other two here.

Yet the style of delivery is just as impressive as the power itself: there is a relentless rush of performance right through the gears, which are much better spaced than the 100 model. Flexibility, too, is excellent; there is little turbo lag as such and sufficient surplus punch for one not to worry about any slight lack in drive below 3,000 revs. But much of the 100's aerodynamically induced economy has gone – the 200 slips easily below 20mpg driven hard and averaged 20.9mpg.

Feeding 182 horsepower through the front wheels is no mean feat and the 200 manages the task pretty well. There is less steering fight than one might fear but sometimes on an uphill corner or powering away from a standstill, the wheel does torque-steer aggressively from side to side. It needs care in the wet, too.

More annoying is the jerkiness inherent in the engine and driveline which can make smooth gear shifting very difficult. The weakness of the five-cylinder engine's character is the way it responds so abruptly to the throttle being floored or lifted, and the flexible mountings needed to absorb these and its internal vibrations create a jerky stop-start style of gear changing that mars the smooth flow of power. The automatic 'box would cure things – but reduce performance, too.

The front-hung engine, right up in the nose of the Audi, gives it an impressive surefootedness on the road. It runs arrow straight, even at the highest speeds, and round corners those fat rubber covers ensure that the nose heaviness does not translate into early understeer.

In fact, the 200 clings on very well indeed and superb power assistance, which gives the steering the weight of a family car's and retains feel of what is going on at the front end, is an immense help in driving quickly along country roads.

With the lower, fatter tyres has come a reduction in ride refinement, however. The suspension is taut and well damped at higher speeds but round town or on poor surfaces, bumps can crash through. There is also a marked increase in road roar – and the only feature spoiling the 100's silent progress at speed has become a serious irritation on certain coarse tarmac or concrete motorways in the 200. Since the turbo engine can be intrusive, too, no longer can one revel in Audi's marvellous high speed quietness – retrograde progess in a more costly model.

Braking is highly effective and on dry roads the car can be slowed rapidly from high speeds without triggering off its ABS system.

Though the Audi is physically the smallest of the group, it doesn't feel so inside where the large, curving glass creates a spacious, light cabin. It's not a particularly impressive one, though, being tidily finished rather than sumptuous: plain cloth seats, and plastic door and dashboard mouldings aren't much to shout about in this company.

But plain as it might be, the seating is actually rather good. The big front seats are comfortable – softer than the German norm – and offer good support. Proper height adjustment would be welcome, though, instead of the halfway house cushion tilt adjuster.

The back seat is truly luxurious, not so much for any extra inches it offers (all three in this group have legroom to spare) but for the relaxing, comfortable shaping of its bench. Sockets for individual radio headphones complete the sense of self-indulgence.

Yet the 200 still cannot help being a polished up 100 to those who have come across both: the dashboard is almost the same – a logical but hardly impressive affair that supplements the bare minimum dials with a useful check panel and a less useful mpg meter (amusing to watch it leap from 3mpg to 99mpg when accelerating then lifting off!). Instead it is by adding a host of luxury fixtures and fittings that Audi have created the 200's upmarket appeal: air conditioning, which though noisy, makes the interior tolerable in summer; electric windows; central locking; ABS brakes and all you need for a radio except the radio itself.

Nice features to have, and the match of any here, but they still don't make the 200 stand out in the car park.

BMW 732i

IT MAY have been around now for more than seven years, but BMW's ultra-conservative attitude to change means that the 7-Series hardly shows its age alongside the company's newer models. It is still a less than inspired piece of styling, however, whose awkward bulky lines are too clearly the result of

an attempt to stretch the outlines of the smaller cars into a big saloon.

But an immaculate finish and gleaming chromework manage to disguise some of its dull looks and create in the observer the desired impression of opulence and style. It's the same when you drive the 732i: the good features manage to promote themselves forcibly and disguise the weaknesses.

The best feature of all has to be the superb engine. The big six is as silky smooth and as quiet as anyone could wish; it revs with the kind of keen, urgent fluency that one normally only expects to find in small capacity, lightweight units.

The automatic gearbox might be expected to rob it of some of its urgency but, as auto-boxes go, BMW's four-speeder has to be rated one of the most responsive and smoothest shifting.

Whether the added complexities of the switchable box are worthwhile is debatable, however: effectively the Sport setting simply cuts out fourth gear entirely and makes part-throttle kick-down sharper. Floor the throttle and the 'box behaves the same way in either mode. After playing with it for a few miles, one tends to leave the box in the Economy setting in order to benefit from the added quietness and economy of the overdrive fourth speed as often as possible.

Whether in Sport or Economy, the 'box responds to a floored throttle by snapping down a gear in a marvellously jerk-free shift, and every up-shift – whether flat out at maximum revs or not – is a smooth blur. In the 1-2-3 mode the transmission can be used just like a clutchless manual 'box with the driver having complete control over the choice of ratio.

In outright performance the 732i is quick, but not exceptionally so, with its blunt body-shape limiting top speed to 129mph and a 0-60mph best of 10.1sec. On the road, though, the smooth punch of the engine allied to the slick dexterity of the transmission mean it is rarely short of zip when the situation demands. The Motronically managed engine has a relatively modest thirt for fuel, too – 21.2mpg overall.

It isn't such a satisfactory car to handle, however. It prefers the open roads and as the bends get tighter its cornering manners get worse. It always feels a big car to manage, that's one reason: bigger than it really is, perhaps, because of the way the seats are set far apart from each other and close to the outside edges of the car.

But size isn't the only reason. On faster curves one is impressed by the weight and accuracy of the steering – recirculating ball though it may be – and the taut stability of its suspension. But turn into tighter corners and one notices too much roll and an over-soft front end that sets the 732 into an ungainly corkscrewing. Apply too much power or none at all, and the rear end will wag untidily, especially in the wet. It's messy handling rather than poor handling for there is a generally high level of grip but one wonders whether the sports suspension pack (no extra charge for this option!) might tidy things up.

It would probably firm the ride up more: it's firm now but only noticeably bumpy at slow speeds on country roads, and has that big car solidity at speed, marred only by some intrusive wind noise.

BMW's pursuit of technical trickery has turned the interior of the 732i into something close to Cape Canaveral's control centre than the elegant home-from-home an Englishman bred on walnut burr Jaguar trim might expect in his big car. The once relatively straightforward dashboard has had check panel, computer (£516), electric front seat adjustment with memory (£840) and air conditioning (add £1,657) shoehorned into it on our car. It all works with impeccable precision, of course, but it can be confusing, and the grey interior trim is rather drab.

The front seats are orthopaedically firm, large and cleverly adjusted by electrics that can even move the headrests up and down. The rear bench is not short of leg or headroom, though the seat itself is flatter, firmer and more upright than might be appreciated on a long run.

You probably have a good idea by now of what is available from BMW's extensive extras list. For those who can't make their minds up, the Special Equipment model at £19,325 throws in ABS brakes, the non-switchable autobox, electric sunroof, alloy wheels and one or two other tricks. Still missing if the BMW was to be put on a par with the others here would be air conditioning. But you will be pleased to know that central locking, electric windows and headlamp washing are all standard.

It could never be called a bargain, but at least you can spend on options you fancy and are not forced to pay for those you don't.

JAGUAR SOVEREIGN

LIKE AN ageing film star trying to retain his matinee idol looks, the Jaguar looks elegant, handsome, dated in some areas and, worst of all, tarted up in others – particularly those dreadful drilled wheels which don't suit it at all.

There are the same mixed feelings when you drive it: admiration that it can still be so good in so many ways, annoyance that so many little faults remain that have been with the car over the years, disappointment that some of its good features have been lost

Far left: Audi's instrumentation looks a bit sparse and includes a nonsensical digital instant mpg readout. The blank switches in the centre are also rather downmarket obscuring the fact that the 200 Turbo is the best equipped of the three with items such as air conditioning and ABS brakes included as standard. Middle: the extras-laden BMW has a bewildering array of switches, lights and buttons but also beautifully presented instruments. Matt wood trim comes nowhere near the opulent luxury of the Jaguar. Below: the polished walnut and leather trim of the Jaguar are superb but the switchgear and reflective instrumentation are pretty old-fashioned

with the passage of time.

The elderly engine comes as a reminder of the Jaguar's origins – the long stroke unit is red lined at a very modest 5,000 revs and, frankly, is starting to sound a bit past its best even before that. But it is a responsive motor and its big capacity ensures that there is always ample torque prompting the car forwards, even with light throttle openings.

It is a heavy car, of course, and while the relaxed way it answers the throttle is impressive, it never really gets into a stride that can match the other two. A very lazy gearbox doesn't help at all: it won't kick down into bottom beyond about 30mph, nor from top into second above some 65mph. The full throttle change points were odd on our car, too – at three-quarter revs out of first and then it would force the poor old engine right through the red line in second, making the driver lift off momentarily to encourage it into top. Most likely needing some adjustment, the shift was also disappointingly jerky compared with past Jaguars and not in the same league as the BMW's.

The Sovereign can be wound up to a maximum speed of 126mph without problems, but by now the tacho needle is creeping into the red zone so this is no overdriven autobahn cruiser, and fuel economy sufers, too, from the weight, gearing and other incipient middle-age factors – 18.6mpg was our overall average.

But if the engine is a sign of the Jaguar's age, its handling and ride certainly aren't. The steering is as lifeless as ever, sadly, if a bit heavier now than in the past. But once one has got used to having no sensation at all of what the front wheels are doing, the big saloon can be managed with assurance.

It is softly sprung, considerably softer than the German pair, but still has an uncanny reassurance and poise to its handling. It remains neutrally balanced right up to its high limits of grip, when a gentle understeer signals the start of progressive breakaway. Mess it about by lifting off or using too much throttle and the Jaguar still won't bite back – that's what comes of having "proper" constant camber rear suspension.

Its softness of springing gives the Jaguar an excellent ride, especially over country roads where it poise really shows through. The only shame is that the P5 tyres promote some rumble and coarseness over certain surfaces which was never there in the past.

Yet it isn't the indulgent ride that creates the Jaguar's luxury: open the door and the smell of leather wafts out of that gentleman's club interior which is probably still the car's biggest sales weapon. Paradoxically, its weakest feature, too.

The low, elegant exterior lines confine interior space, and headroom is not over-generous either front or rear. The driver finds himself sitting low in the car, despite a rather puny cushion tilt adjuster, and the seat is too short and soft for real support under the thighs.

In front of him is a muddle of a dashboard that has been updated too often for its own good. The walnut veneer is lovely and there is a sensible number of dials, but they are cheap looking and so, too, are the switches whose awful shapes and graphics remind one of British cookers of the '70s. Long standing irritations remain; like the feeble, jerky action wipers and the steering column that adjusts for reach but leaves the stalks behind, out of reach.

Oddly, for a luxury car, the Jaguar's back seat remains one of its worst points: the backrest is too upright and the short, flat cushion gives little thigh support. The boot, too, is small and shallow compared with the whale-like proportions of the BMW and Audi holds.

But there is no shortage of luxury in the Sovereign: leather trim, air conditioning, stereo cassette system, central locking, electric windows – even ankle deep woolly rugs for the floor are all part of the Sovereign specification – yet it is still the walnut and leather that leave the lasting impression.

Top: Audi's turbocharged staight-five sits way up front and makes the 200 Turbo the quickest, if not the most refined, of the trio. Middle: BMW's silky smooth six is the pick of the bunch. Bottom: the old XK engine powers the heavy Jaguar along at a respectable pace but is rather thirsty

CONCLUSIONS

THE PERSON buying one of these luxury carriages is buying more than a car; he – or she – is choosing status as well. And that is what the Audi lacks; it is a fine, fast, senior executive's car that is, arguably, best of all three when it comes to carrying four adults long distances in speed and comfort. But it's not ostentatious, nor does it offer any rational alternative to the gleaming chromework of traditionalist luxury. Just the likelihood, instead, of being confused with a cheaper model.

That may fit in well with the thinking of large corporations, who want to reward achievement in their senior staff but restrain individual expressions of personality. However, we can't really see the 200 appealing to the sort of self-made, self-promoting businessmen who so often buy BMWs.

What might appeal to them is the Jaguar, going out in rare style with record sales in this its last full year. It's a beautiful looking car still; low and lithe compared with the dreary slab sided BMW and it reeks – literally – of luxury. Certainly it shows its age in many ways: the poor fuel consumption, relative lack of interior space and muddled facia, but it still shines in many others – handling, ride and sheer refinement are unsurpassed here.

The BMW (or a Mercedes) has been the instinctive choice in this market for so long, while Jaguar has been in the doldrums. Indeed, logic would suggest that the 732 still is the wise buy: relatively good economy, electronic monitoring of service requirements, ABS braking and so on. Yet the impression remains that the 732i is rather a dull car, relieved by an impressive engine and transmission, flattered by technological tricks.

That the Jaguar can outsmart it in so many ways must rankle with the Bavarians. We would be swayed – just – by the Jaguar's elegance rather than the BMW's logic. But the real question for both companies is what Jaguar will reveal in the new XJ40. Can a car so good that its looks, its handling, its ride and its luxurious quality still beat the best after so many years, be replaced by one that retains all these strengths and cures some weaknesses? If it can – BMW look out!

GENERAL DATA

	Audi 200 Turbo	BMW 732i auto	Jaguar Sovereign 4.2 auto
Price	£17,013	£17,450	£18,995

ENGINE

Cylinders	5 in-line	6 in-line	6 in-line
Capacity, cc	2,144	3,210	4,235
Bore/stroke, mm	79.5/86.4	89/86	92.1/106
Valves	Sohc	Sohc	Dohc
Compression ratio	8.8:1	10.0:1	8.7:1
Fuel system	Turbocharger/fuel injection	Bosch LE Jetronic	Lucas/Bosche injection
Max power, bhp/rpm	182/5,700	197/5,500	205/5,000
Max torque, lb ft/rpm	186/3,600	210/4,300	236/3,750

TRANSMISSION

Type	Fwd, 5-speed manual	Rwd, 4-speed auto	Rwd, 3-speed auto
Internal ratios and mph/1,000rpm			
Fifth	0.78:1/23.5	—	—
Fourth	0.97:1/18.8	0.73:1/30.4	—
Third	1.36:1/13.5	1.00:1/22.2	1.00:1/24.8
Second	2.12:1/8.6	1.48:1/15.0	1.45:1/16.9
First	3.60:1/5.1	2.48:1/8.9	2.39:1/10.3
Final drive ratio	3.89:1	3.45:1	3.07:1

SUSPENSION, STEERING, BRAKES

Front	Independent by MacPherson struts; coil springs; anti-roll bar	Independent by MacPherson struts; coil springs; anti-roll bar	Independent by double wishbones; coil springs; anti-roll bar
Rear	Torsion beam axle located by trailing arms and Panhard rod; coil springs; anti-roll bar	Independent by semi-trailing arms; coil springs	Independent by lower wishbones and fixed-length driveshafts; coil springs
Steering	Assisted rack and pinion	Assisted recirculating ball	Assisted rack and pinion
Brakes, front/rear	Ventilated discs/discs	Ventilated discs/discs	Discs/discs

WHEELS, TYRES

Wheels	Alloy, 6J × 15	Steel, 6½J × 14	Alloy, 6J × 15
Tyres	205/60 VR 15	205/70 VR 14	205/70 VR 15

DIMENSIONS

Wheelbase, in	106	110	113
Front track, in	57.8	60	58.5
Rear track, in	57.8	59.4	58
Overall length, in	189.2	191.5	195
Overall width, in	71.4	70.9	69.8
Overall height, in	56.0	56.3	54
Tank capacity, gall	17.6	22.4	20
Kerb weight, cwt	25.4	29.3	36

PERFORMANCE

Maximum speed, mph	141	129	126
Speeds in gears	(at 6,500rpm)	(at 6,200rpm)	(at 5,000rpm)
First	33	55	51
Second	56	93	84
Third	88	—	—
Fourth	123	—	—
Acceleration through the gears, sec			
0-30mph	2.9	3.6	3.9
0-60mph	8.5	10.1	10.6
0-100mph	21.5	28.5	29.1
Acceleration in fourth (kickdown for BWM and Jaguar), sec			
30-50mph	7.6	3.4	4.2
50-70mph	6.5	5.7	5.8
70-90mph	7.1	9.0	8.8
Acceleration in fifth, sec			
30-50mph	11.9	—	—
50-70mph	9.4	—	—
70-90mph	10.4	—	—

FUEL CONSUMPTION

Overall test, mpg	20.9	21.2	18.6
Likely variation, mpg	18-24	19-26	16-21
Government figures, mpg			
Urban cycle	23.0	17.5	14.5
Steady 56mph	39.8	36.2	28.2
Steady 75mph	31.4	28.7	23.7
Maker	**Audi NSU, W Germany In UK: VAG (United Kingdom) Ltd, Yeomans Drive, Milton Keynes MK14 5AN**	**BMW, W Germany In UK: BMW (GB) Ltd, Ellesfield Avenue, Bracknell RG12 4TA**	**Jaguar Cars Ltd, Browns Lane, Coventry CV5 9DR**

Performance tests carried out at Millbrook Test Track, Bedfordshire.

JAGUAR XJ6

JOY OF JAGUAR'S CLASS

Following the Jaguar XJ6 launch in 1968, substantial advances have been made. Graham Robson takes a look at the viability of the post-1979 Series III on the secondhand market.

Five years ago, the Jaguar marque was in trouble. The quality reputation was in tatters, and exports to the USA were being hit very hard by the strength of sterling. Then came the *'Boy's Own Paper'* miracle. A new chief executive arrived, to spearhead a blitz on poor reliability, the £/$ parity began to move in a favourable direction — and the Series III cars made their mark.

The miracle is now complete. The XJ6 series is selling faster than ever before, so much so that the launch of a replacement model, code-numbered XJ40, has been put off repeatedly.

Autocar has previously surveyed the XJ6 models in the *Buying Secondhand* series, but never concentrated on the Series III models, for which sales began early in 1979. There are so many of them around now, however, that a very representative check-out can be made, especially as we also cover the near-identical examples carrying Daimler badges and radiator grilles. Nearly 100,000 six-cylinder engined Series III saloons have now been built, of which at least half have been delivered in this country.

BACKGROUND

There is, of course, a good choice, but the principal opposition comes from the 7-series BMWs and S-Class Mercedes-Benz models. It is a hotly-contested sector, where depreciation is high, and 'bargains' are difficult to find.

The original Jaguar XJ6 was launched in 1968, and the current Series III car is a close, lineal, descendant, of that design. The Series II car, in fact, was put on sale late in 1973, and the Series III took over in 1979. The body shell is still essentially the same as always, except for a different glass 'greenhouse' and decorative details, and all have been built on the same 9ft, 4.75in wheelbase.

All cars in this series have independent front and rear suspension, plus four-wheel disc brakes. There have been two sizes of the famous twin-overhead-cam XK six-cylinder engine (3.4-litres and 4.2-litres), and a choice of manual or automatic transmission. Every car has power-assisted rack and pinion steering and all the 4.2-litre engined versions are capable of at least 120 mph.

They are big, and heavy cars — as the overall length of 16ft 3ins and unladen weight figure of 3,900lb confirm. Fuel economy is not a strong point (the original car was designed at a time when fuel prices were low) — about 18 mpg being normal.

ENGINES

Although we do not cover them in this survey, it is well to remember that there is a V12, 5.3-litre version of this design (now called the Sovereign HE), and that the XJ-S Coupe/Cabriolet model uses the same (shorter-wheelbase) type of underpan, and the same suspensions.

The Series III Jaguar XJ6 superseded the Series II in 1979. All the cars in the series use a version of the twin-cam 1958 XK engine, with several capacity options. The exterior remained very similar but the interior options were considerably uprated; front seats with adjustable lumbar support, stereo/radio cassette, electrically operated sunroof, electrically controlled door mirrors

All the cars covered in this survey use a version of the twin-cam XK engine, first seen in 1948, and gradually developed ever since. By modern standards it is neither powerful, economical, nor high-revving, but it is silky-smooth unless over-revved.

The vast majority of all Series III XJ6 cars use the 4.2-litre version, which has Bosch/Lucas fuel injection, and a peak power output of 205 bhp at 5,000 rpm, with strong and impressive torque from very low speeds indeed.

There has always been a 3.4-litre option of this same engine design (but never behind the Daimler radiator grille and badge), and this is rated at 161 bhp (DIN) at 5,000 rpm; induction is by twin SU carburettors. Only about 5,000 of the first 100,000 Series III cars were so equipped.

TRANSMISSIONS

The vast majority of all Series III cars have been built with Borg Warner automatic transmission (Model 65 at first, Model 66 from summer 1979). The option, on all except the Daimler Vanden Plas, was the latest Rover-Triumph five-speed all-synchromesh manual transmission (as used in the Rover 3500, Vitesse, and Triumph TR7/TR8, for instance), which incorporated an 'overdrive' fifth gear, with a direct fourth ratio. For those who still liked to shift the gears, this was a very handy box which also offered better prospects for fuel economy. There was no price difference between manual or automatic transmission cars.

BADGES, TRIM PACK, OPTIONS

In one respect, at least, this is simple enough to sort out; all cars, of whatever type, have been built on the same basic body shell/monocoque structure, which is itself four inches longer in wheelbase (and rear leg room) than the original of the late '60s. All cars have four doors, and there has never been a limousine, or division, option.

Compared with the superseded Series II cars, now there were front seats with adjustable lumbar support, a stereo radio/cassette installation, quartz-halogen headlamps, and a conventional two-speed wiper mechanism. Important extras included an electrically operated sunroof, electrically controlled door mirrors, cruise control, headlamp wash/wipe, electric driver's seat height adjustment (this was standard on the VDP, along with wash/wipe and electric mirrors). Other extras included full air-conditioning, inertia reel rear belts, and XJ-S type alloy wheels.

The initial line-up was Jaguar XJ6 3.4-litre and 4.2-litre, Daimler Sovereign 4.2-litre, and Daimler Vanden Plas 4.2-litre, and this continued unchanged for the first year and a half. For 1981, however, the specification of the 3.4-litre car was slightly cut▶

Left: The XK engine has been in production for 35 years, and still goes wrong as it gets older. Check the oil pressure, look for oil leaks and for a smoky exhaust. Right: The reliability of the suspension is good, despite being complicated. Damper changes are recommended every 30,000 miles. It is very important to check the suspension alignment, as damage to tyres can be very costly

PARTS PRICES

	3.4-litre	4.2-litre
Engine assembly — bare (exchange)	£1,290.30	£1,351.25
Gearbox assembly (new)	£923.45	£923.45
Clutch driven plate (new)	£45.60	£45.60
Clutch, complete (new)	£105.80	£105.80
Automatic transmission, with convertor (exchange)	£461.15	£461.15
Brake pads — front set (new)	£24.55	£24.55
Brake pads — rear set	£14.65	£14.65
Suspension dampers — front (each)	£25.88	£25.88
Suspension dampers — rear (each)	£31.45	£31.45
Water radiator assembly (new/exchange)	£155.25	£155.25
Tyre price, typical	£136.16	£118.80
Alternator (new)	£37.09*	£137.38
Starter motor (new)	£105.80	£105.80
Headlamp unit — main (each)	£40.54	£40.54
Taillamp unit — (each)	£56.93	£56.93
Front wing panel	£194.35	£194.35
Front door — skin panel	£86.83	£86.83
Bumper, front, complete (new)	£284.05	£284.05
Bumper, rear, complete (new)	£276.92	£277.92
Windscreen, laminated	£142.60	£142.60
Exhaust system — main silencer box only	£29.10	£29.10
Exhaust system complete	£185.38	£185.38

**Exchange* *All the above prices include VAT at 15 per cent*

◀ (cheaper radio/aerial equipment accounted for much of this) and the price was reduced by £500. The 3.4, incidentally, had central locking, and cloth upholstery by this time. In the first three years, several previously optional items were also standardised on the 4.2-litre models.

There were significant changes for 1983 (commensurate with the dropping of the 'Daimler' badging for European markets), but no mechanical upheavals. The facia was somewhat restyled, to include a Lucas trip computer, and controls were re-positioned, while the once standard 'attaché case' tool kit (discarded years previously) had been reinstated.

In the autumn of 1983, and still current, came the final changes. The Daimler version lost its 'Sovereign' and 'Vanden Plas' badging, and became plain '4.2-litre', there being only one version. The Jaguar line-up became 3.4-litre, 4.2-litre, and Sovereign 4.2-litre. Automatic transmission was an optional extra, still, on 3.4/4.2 Jaguars, but a no-cost feature on other types. The Jaguar Sovereign had cast-alloy wheels, and all previous Vanden Plas/Daimler Sovereign features, as standard.

AVAILABILITY

If you consider XJ6s of all ages, then there is no shortage on the secondhand market, for the earlier models are well-and-truly in the 'rusty fast banger' category by this stage. Indeed, among some sectors of the community you may find more middle aged Jaguars and Rover SD1s than Cortinas. The Series III cars, however, are in more limited supply, and still cost too much, at two or three years old, to sell quickly.

The rarest derivatives are the 3.4-litre models, which only made up less than five per cent of Jaguar's output. We tested a Series II car some years ago, and found that it had a 117 mph top speed, but acceleration and fuel consumption that could be bettered by an injection Granada.

Daimlers are far less common than Jaguars — perhaps a one to three ratio — though at three years old they are still worth up to £800 more than the equivalent Jaguar. If the Daimler has Vanden Plas badging and trim, add another £1,000 to the price.

Manual transmission cars are rare, even though there was nothing at all advanced about the direct-top automatic transmission used instead. The majority of XJ6 buyers, it seems, preferred to let the automatic do the driving (or the chauffeur, in quite a number of cases). Manual transmission cars, therefore, are a rare treat, which our testers enjoyed driving. Technical editor Michael Scarlett even completed Land's End to John O'Groats, successfully, entirely in fifth gear, not too long ago!

For value, therefore, we're sure that a manual transmission 4.2-litre Jaguar XJ6 is the best bargain; we really cannot see the point in spending all that extra money on different badging, or yet more plush on what was already a well-equipped car.

At the moment, incidentally, there are 350 Jaguar-Daimler dealers in the UK, and since the cars remain in full production, parts and service expertise are freely available. One word of caution, though: Although you can buy a three-year-old car for well under £8,000, the parts are priced in line with new car prices — £18,995 for a Sovereign 4.2. Can you afford it?

WHAT TO LOOK FOR

First of all, take your time to track down a car with the specification and badging you need. There is really no need to rush in to buy a car if it has the Vanden Plas gloss (if, that is, you don't want it), for there are always plenty of these cars on the market. Many used Jaguars tend to return to the garages which originally sold them, or to BMW, Mercedes-Benz or perhaps Ford outlets where a conquest sale has been made. These days, it seems, the 'conquests' are the other way around.

First owners of XJ6s were almost invariably companies, or self-employed professional people and they would usually be well serviced. On the other hand, they seem to change hands quite early, and from about the third year onwards, care and maintenance seems to slip somewhat. So have a good look at the service record, and try to get a feel for the type of motoring the car has already had.

BODY AND TRIM

There are two points to make immediately. The oldest Series III car is still less than six years old (most are two to four years old, on the secondhand market right now), and Jaguar quality has improved considerably in recent years. It follows that you would be very unlucky to find much rust corrosion in evidence as yet.

Obvious neglect will show up in rusting around panel seal lines, the sharp edges of the front panels (and where various lamps fit into the bodywork), and along the bottom edges of doors, the edges of the wheel arches, and the boot lid. There should not have been time for the tops of the front wings to be affected, or for other upper panels, though the sills behind the wheel splash areas may already be in trouble from stone chipping. XJ6 wheels stand well out towards the edges of the car, so this problem is more severe than on their contemporaries.

Although paint quality and underbody treatment on the Jaguar XJ6 is now much better than it was on the car in the early '70s, it is still worth looking under the car for trouble.

This is especially important if there is any question of the car having been involved in an accident at some stage in its earlier life. Look, particularly, at the condition and alignment of the structural side members and

APPROXIMATE PRICES

Price range	Jaguar XJ6 3.4-litre	Jaguar XJ6 4.2-litre	Jaguar XJ6 Sovereign 4.2-litre	Daimler Sovereign 4.2-litre	Daimler Sovereign Vanden Plas 4.2
£3,400-£3,600	1979	1979			
£3,800-£4,000				1979	
£4,400-£4,600	1980				
£5,300-£5,500		1980			1979
£6,000-£6,200				1980	
£6,300-£6,500	1981				
£7,400-£7,700		1981			1980
£8,200-£8,600	1982			1981	
£9,500-£9,800					1981
£10,200-£10,600		1982			
£10,700-£11,000	1983			1982	
£12,500-£12,800		1983			
£13,000-£13,400				1983	1982
£14,500-£14,800		1984			
£16,000-£16,300					1983
£16,500-£16,800			1984		

Note: All values quoted are for cars with automatic transmission. Cars with the five-speed manual transmission are worth about £500 (1979 models) to £1,400 (1983 models) less than the automatic transmission cars.

SPECIFICATION AND PERFORMANCE

	XJ6 4.2-litre Automatic	XJ6 4.2-litre 5-speed manual
Specification:		
Engine size (cc)	4,235	4,235
Engine layout	Twin-cam six	Twin-cam six
Engine power (DIN bhp)	205	205
Car length	16ft 3.3in	16ft 3.3in
width	5ft 9.7in	5ft 9.7in
height	4ft 6.1in	4ft 6.1in
Boot capacity (cu ft)	19.1	19.1
Turning circle (kerbs)	39ft 9in approx	39ft 9in approx
Unladen weight (lb)	3,875	3,871
Max payload (lb)	900	900
Performance summary:		
Tested in *Autocar*:	29 Dec 1979	7 Jan 1984
Top speed (mph)	127	131
0-60mph (sec)	10.0	8.6
Overall fuel (mpg)	16.8	18.3

the inner sills, and check the pick-up points for rear suspension radius arms, and of the front cross-member/sub-frame.

An XJ6 seems to suffer more than its share of traffic scrapes (that shape, once again), so look carefully at all the corners, and exposed edges. Look, too, at the condition of the leather seat facings and wood facia and door cappings; these are surprisingly costly to replace.

Heating and ventilation systems often seem to go wrong, in detail, and by leaking, rather than fundamentally, so check out the performance during a test run. There are hordes of electrical fittings, and a complete 'cockpit check' is called for.

MECHANICAL

Let's start with the engine. Even though the XK engine has been in production for 35 years, it can, and does, still go wrong as it gets older. On the test run, be sure the oil pressure (up to 40 to 50 psi is right) is correct, and that the exhaust is not smoky; after the test run is there any sign of oil leakage from the engine (look underneath, after a few minutes). Have a look at the condition of the engine cooling water after the run, to be sure there is no oil contamination, for these engines sometimes suffer cracked cylinder blocks. In the same inspection, be sure there is a permanent inhibitor in the water, for there is an aluminium head.

A good engine should last up to 100,000 miles without needing major attention, and it will begin to show old age in terms of noisy valve gear, and a clattery timing chain. Rebuilding them takes time, and costs a lot of money; you can't really tackle it yourself. The fuel injection system, as far as we can see, seems to be reliable, and the economy good if well set-up.

The Borg Warner automatic transmission doesn't have too much torque capacity in hand to deal with this engine, so be sure that it is still operating crisply, without sluggish changes, and that there are no leaks. Check out the operation of the lever quadrant — even its feel (is it gritty? It shouldn't be). Manual transmissions may show old age with noise in the lowest gears, and worn synchromesh, particularly second, and be sure that the clutch is still doing its job, particularly when you most need the punch. You may find a rather 'clunky' drive line, for there are a number of universal joints. Look particularly at the prop-shaft centre bearing, and the rear drive shafts.

There is quite a lot of suspension complication, front and rear, but in general, reliability is good. The makers recommend damper changes at 30,000 miles, which we think is disgracefully low — but, is the ride floaty or well-controlled? Are there any creaks, on-throttle/off-throttle clunks at the rear, or any semblance of steering instability?

At the front, quite a number of XJ6 power steerings whistle and squeak, but the piping shouldn't leak. After two or three years' hard use, expect bush and ball joint wear in the suspension, easily if not cheaply rectified. Brake discs, especially at the front, may have distorted.

MODELS AVAILABLE

March 1979: Series III Jaguar XJ saloons announced, with Jaguar and Daimler badges, twin-cam 'six' and vee-12 engines. All on same wheelbase, with same four-door style. 6-cyl models were available as follows:

Jaguar XJ6 3.4-litre, Jaguar XJ6 4.2-litre, Daimler Sovereign 4.2-litre, Daimler Vanden-Plas 4.2-litre.

Autumn 1983: Minor badging and specification changes, so that line-up now reads:

Jaguar XJ6 3.4-litre, Jaguar XJ6 4.2-litre, Jaguar Sovereign 4.2, Daimler 4.2-litre.

January 1985: The above range continues in current production.

PRODUCTION TOTALS

Production still goes ahead, at a steadily rising rate. At the time of writing, however, *approximate* production totals for each major type are:

Jaguar XJ6 3.4-litre 5,000
Jaguar XJ6 4.2-litre + Sovereign 90,000
(including Daimler-badged cars)
— more than 100,000 of the Series III saloons (vee-12 engines included) have now been built, and a total of more than 26,000 were built in 1984.

Be absolutely sure the suspension alignment is correct; check the state of the tyres, and the trend in tread wear, for an XJ6's rubber is very costly to replace. Worn suspension bushes may affect precision and stability. While you are under the car, too, ask someone to start up the engine for you to inspect the exhaust system, which may not only be corroding (three years, at best, we'd say), but may be fouling the underside and body shell.

Finally, please remember this. The XJ6 may ride better, and be at least as quiet as, a modern Rolls-Royce, but it certainly doesn't last as long. Even though you may have to pay less than £10,000 for a 30,000 mile car, have you considered all the running costs? You should. ■

THE SPECIALTY FILE

DMJ Jaguar XJ6

At least one guy knows what he sees in the big black cat.

PHOTOGRAPHY: DICK KELLEY

• Ann Arbor is awash with U of M students working part-time jobs, some of them at our favorite automobile-laundering facility. Among those who labor there is a black psychology major who has established a mental tabulation of our zanier test cars. When the black DMJ Jaguar XJ6 came glowering out all adrip with sparkle, the kid lingered long with his terry towel, caressing the swoop of the car's flanks with his eyes and carefully wiping the last potential water spots from the sheen of its midnight-glossy skin. Then he chuckled and said, "This car is *me!*"

Who were we to argue?

Except for its shiny alloy rims and tan leather insides, the DMJ XJ6 could rival the Toy Store East's Supra Sarizer (*C/D,* October) as the blackest car in memory and one of the most striking ever to make our car-wash tabulation (although which side of striking the DMJ falls on is obviously in the eye of the beholder). If the delivery of stares, smiles, and oohs and ahs is what you want, the DMJ XJ6 can handle it. Unabashed stares, king-size smiles, and high-decibel oohs and ahs have been stock in trade among wealthy Texans for years. Those provided by the DMJ have been dreamed up and applied by an interesting if unlikely pair of Texans named Kirsten Dodge and Joe Maxwell. The initials of their last names provide the first two letters of their first car's name (the last initial standing for Jaguar itself), and their approach to the specialty-car business is new and different, as they are.

Ms. Dodge is a striking woman, with a business background, a long attachment to horses, and an abiding interest in Jaguars. Mr. Maxwell is a former farrier and construction specialist whose father was a fully distended Jaguar fanatic. Maxwell and Dodge met when he was doing pedicures on her family's horses, but the subject of big British cats soon arose.

Dodge says, "After some involvement with earlier Jaguars, we started buying used XJ6s that we could get for good prices and recondition. We found out that there was a whole lot more that we wanted to do to make the car more exciting and more individualized. About two years ago, Joe came up with the idea of the DMJ XJ6. I was actually rather stunned about the scope of it, but we went ahead anyway."

Maxwell says, "The plan is to build around 50 of these cars. First this model, and then later we'll do others. We're not there yet, but we're getting close. Right now we're split, with half the shop in England and half in Texas, near Austin.

"We bought a number of Jaguars, and I systematically stripped them apart to see where the strong and weak areas were, and then we rebuilt them to my specifications from the shell up. We can build them from new or used to almost any configuration a customer might want. This car has Recaro C seats, full leather upholstery, and a four-spoked, leather-bound steering wheel. We can provide six- or twelve-cylinder engines in a wide band of outputs. We can also replace the original engines with small-block Chevrolet V-8s.

"The engine you have here is a European six-cylinder fitted with a five-speed transmission and U.S. emissions controls."

In this trim, the 4137-pound Jaguar whips up 0-to-60-mph times of 9.6 seconds and a top speed of 122 mph, 2.1 seconds and 11 mph better than the last XJ6 we tested (a stock 176-bhp, three-speed automatic 4020-pounder, featured in the November 1981 issue). Most of the DMJ's 117-pound weight gain comes from its prototype steel air dam, side skirts, and rear wing, which will be made of aluminum (not fiberglass!) on the official run of cars. All of these pieces, be they steel or aluminum, are a source of heated controversy even among hot-sedan lovers, who either adore or despise the Dodge-Maxwell alterations to Jaguar's classic shape. The DMJ's rounded whale tail rouses the most emotional flap of all, though its aerodynamic effect does seem to provide a useful gain in high-speed predictability.

There are, however, several small flies buzzing their little wings off in the $60,000 DMJ's ointment. The drivetrain vibrates noticeably over 4000 rpm, the shifter is set to the right, and the clutch is heavy—all of which add to the DMJ's racer feel but move it a good distance from traditional Jaguar silkiness. Tradition has also been snubbed in the wheelhousings, where Pirelli P7s must deal with a newly firmed suspension, strong understeer, and more weight than they want to know about (at 32 psi they howl plenty). This combination results in a lowly 0.70 g on the skidpad and 206-foot stops from 70 mph and puts the DMJ XJ6 no more than even with a stock XJ6 in test-track performance.

Although Maxwell admits he's new at the chassis-tuning business, he rejoices in a special winding road near the Austin shop, and he claims DMJ can tune ride and handling to almost any level a customer could want. In the case of our DMJ XJ6, it was the level Maxwell wanted.

Given time, DMJ will probably turn out Jags that handle the way they look. As they say at the carwash, that's "*bad!*" In other words, better.

—Larry Griffin

Vehicle type: front-engine, rear-wheel-drive, 5-passenger, 4-door sedan
Price as tested: $60,000 (base price: $31,100)
Engine type: 6-in-line, iron block and aluminum head, Bosch-Lucas L-Jetronic fuel injection

Displacement	258 cu in, 4231cc
Power (SAE net)	200 bhp @ 5000 rpm
Transmission	5-speed
Wheelbase	112.8 in
Length	199.6 in
Curb weight	4137 lbs
Zero to 60 mph	9.6 sec
Zero to 100 mph	28.2 sec
Standing ¼-mile	17.0 sec @ 82 mph
Top speed	122 mph
Braking, 70–0 mph	206 ft
Roadholding, 282-ft-dia skidpad	0.70 g
C/D observed fuel economy	14 mpg

Nine lives

Photography by Kent Mears/Kevin Brown

Like the cat that couldn't be killed, the Jaguar XJ-6 has lived on beyond its life expectancy. And even now, with a replacement model on the way, this elderly feline is healthier than ever.

by David Robertson

THE JAGUAR XJ-6 should, by now, be a superseded model. It should have been replaced by the long-awaited and much-heralded XJ40. It isn't, of course, and just when it will be is still anybody's guess. However, the third quarter of next year is firming up as a likely launch period.

The incredible fact is that the existing model is selling better than it ever has. Sales have quadrupled in the US in the last three years and in Australia the three models – the $49,900 4.2 saloon, $53,200 Sovereign and $62,600 Vanden Plas – are selling at record levels.

Jaguar engineers are on the record as saying they will not sign off the new model until they are convinced it is "right". It's been thoroughly tested, including a three month stint in Australia. But reports from the UK suggest there are still problems to be ironed out in areas including engine electronics, noise suppression and the availability of its new automatic transmission.

But it is the renewed success of the existing cars that is the more remarkable aspect of Jaguar Cars today. From the time word spread about the re-vitalisation at Browns Lane, Coventry – inspired by the appointment of John Egan as managing director – to the floating of the company on the stock exchange early last year, sales have perked up dramatically.

It is almost as if affluent buyers had been praying for a viable alternative to Mercedes-Benz and BMW.

There's no doubt that the latest examples of the Big Cat are far better built, far more reliable cars than before. And the XJ-6's shape is a timeless styling classic, as smooth and attractive today as it was when the series began in the UK way back in 1968.

They are not perfect, and one still hears the odd horror story. But similar flaws crop up in other luxury cars. No other make offers as comprehensive a warranty cover as importer and distributor JRA's Mastercare, which provides full parts and labour warranty for two years or 40,000 km, as well as paying for all labour and genuine replacement parts used in regular scheduled servicing for three years or up to 40,000 km.

A sweet running XJ-6 is still one of the most pleasant cars to drive with abundant performance from its twin overhead camshaft, 4.2-litre, in-line six and a ride and handling standard that is difficult to match.

The six traces its ancestory back to the famed XK engine of 1949 that stunned critics of the day with performance that included an officially timed maximum speed (in the XK120 sports car) of 212 km/h.

Of course it has been substantially improved with many of its parts made now from aluminium (to reduce weight and improve thermal efficiency), modern hemispherical combustion chamber design and electronic fuel injection. Smoothness is its hallmark and few will argue with its ability to lug 1830kg from rest to 100 km/h in under 12 seconds.

The undersquare (92.1mm × 106mm) cast iron block engine develops 153 kW power at 5000 rpm and 314 Nm torque at a low 1500 rpm. It is a superbly quiet powerplant with good flexibility accentuated by the standard three-speed automatic's sensible ratios and jerk-free changes.

The Jaguar's greatest claim to fame is its supple ride. The suspension package

has always been the benchmark by which others are judged. Maintaining that reputation is another of the reasons why the XJ40 is still under development. It is obviously proving more difficult to retain the XJ-6's standards in the new car, which weighs in at around 200 kg less.

The XJ-6's front suspension comprises fully independent semi-trailing wishbones and coil springs with Girling Monotube hydraulic dampers and an anti-roll bar installed to provide an effective anti-dive geometry. At the rear lower transverse wishbones are used with the drive shafts acting as upper links. Radius arms, twin coil springs (with their own damper units) and Girling Monotube dampers complete the package.

The Big Cat glides along on bitumen in true limousine fashion, insulating occupants magnificently. The suspension shrugs aside potholes and corrugations and, pointed over gravel, displays admirable stability and control.

Its ride and high speed stability are improved by a long 2865mm wheelbase, wide (1480mm front and 1490mm rear) tracks and fat 205/70 VR 15 steel belt radials. Its power assisted rack and pinion steering, while a shade on the light side, bestows great precision and is superbly damped against road shocks.

The vast majority of Jaguar Sovereigns sold locally are fitted with the optional Special Equipment Pack of alloy

The XJ-6 abounds with luxury touches such as power windows and electric outside mirror controls and the entire cabin oozes quality and craftsmanship. The large centre console creates cockpit-like feeling, and the analogue instruments are easy to read.

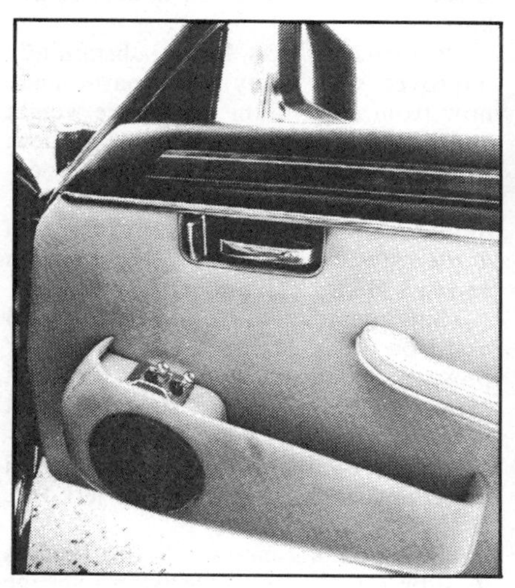

road wheels, cruise control, trip computer and electric sunroof. It adds $3000 to the price.

At high speed the Jaguar is very quiet, apart from some minor wind noise around the door pillars, and its loping ability makes a mockery of 100 km/h speed limits.

The brakes — powered four wheel discs, ventilated at the front — are excellent, acting with a light and progressive pedal and resisting lock up and fade.

It's not perfect. The boot is far too small and head room is lacking in the rear.

The cabin epitomises the English "Old School" of car building with its highly polished burr walnut fascia and stitched Connolly hide upholstery. Here, perhaps, is the true appeal of the Jaguar, for I've seen the workshop where the dashboards are hand-made and the leather chosen and fashioned. There's nothing left in the world, apart from Rolls-Royce, to compare with such craftsmanship.

The materials — they touch the senses with their appearance, smell and feel — give the Big Cat a true luxury character and simply ooze quality.

The driving position is fairly low and the driver and front passenger feel as if they are in a cockpit because the deep centre console sort of encloses them in individual cabins. Controls and analogue instruments are all close at hand and easy to see and read.

According to official AS2077 test results the Jaguar XJ-6 4.2 automatic returns 15 litres per 100 kms on the city cycle and 10.5 litres per 100 kms on the highway cycle although we didn't manage to achieve anywhere near these figures in the real world. With a mixture of conditions, but mainly city and suburban work, we varied between 16.9 L/100 and 20.1 L/100, with an overall average of 18.1 L/100. The Jag is unique with its twin fuel tanks providing 100 litres capacity which should give a minimum range of 500 km and perhaps as much as 700 or 800-odd km on an interstate run.

Luxury abounds from the dial-a-temperature, thermostatically-controlled integrated air conditioning system to electric window lifters and door mirrors, central locking and a quality AM/FM sound system with four speakers and automatic antenna.

The XJ-6 has an image that, amazingly, communicates prestige better in 1985 than it did in 1968. The new XJ40 has a very hard act to follow. □

Counterpoint

A LIFETIME, or the bulk thereof, of reading motoring magazines and watching the cars go by generates in one a love of tradition, style and elegance in cars. And Jaguar is a make that has all of those attributes. It is hard for an enthusiast not to love the Jaguar XJ-6 — because of, and in spite of, its obvious maturity. Burr walnut and leather combine with vintage instruments and controls to enhance this car's charm. But the ergonomics leave something to be desired: the steering wheel is too high and the fore-and-aft adjustment does nothing to help the situation; and the brake pedal is too high relative to the accelerator pedal, increasing reaction time and the risk of not getting the pedal at all in an emergency. The car needs too many different keys, making the owner look like a jail warder on his way home from work; the windscreen wipers wouldn't look out of place on Noah's favourite transport; and the periodic whoosh of air from the automatic climate control when on a very cool setting on a hot day drowns out the sound of the excellent sound system. Opening the sunroof fully creates a drumming at 70 km/h that is unbearably painful to the ears.

The XJ-6 has character and charm but it also has faults and can't sell forever despite them; the XJ40 will need to be a much more modern car to survive. But if it can combine modernity with the grace and style of the XJ-6 it should be a winner. *— Barry Lake*

JAGUAR XJ-6 VANDEN PLAS
4.2-litre, three-speed automatic

ENGINE

Location	Front
Cylinders/cooling	Six, in-line/water cooled
Bore x Stroke	92.1 x 106.0 mm
Capacity	4235 cm³
Induction	Lucas electronic fuel injection
Compression Ratio	8.7 to 1
Fuel Pump	Electric
Valve Gear	Chain-driven twin OHC
Claimed Power	152.9 kW at 5000 rpm
Claimed Torque	314 Nm at 1500 rpm
Maximum Recommended Engine Speed	5000 rpm
Specific Power Output	36.1 kW/litre

TRANSMISSION

Type	Three-speed automatic
Driving Wheels	Rear
Clutch	None

Gearbox ratios

Gear	Ratio	km/h 1000 rpm	Max Speed	At (rpm)
First	2.39	15.38	77	5000
Second	1.45	25.36	127	5000
Third	1.00	36.77	184	5000
Final-Drive Ratio				3.31 to 1

SUSPENSION

Front	Independent by semi-trailing wishbones with coil springs and anti-roll bar
Rear	Independent by lower wishbones and fixed-length driveshafts located by trailing links with coil springs
Wheels	Alloy 6.0JJ x 15
Tyres	Pirelli P5 205/70 VR15

BRAKES

Front	284 mm ventilated discs
Rear	263 mm discs

STEERING

Type	Power-assisted rack and pinion
Turns, Lock to Lock	3.3
Ratio	17.6 to 1
Turning Circle	12.2 metres

DIMENSIONS AND WEIGHT

Wheelbase	2865 mm
Front Track	1480 mm
Rear Track	1490 mm
Overall Length	4959 mm
Overall Width	1770 mm
Overall Height	1377 mm
Ground Clearance	178 mm
Kerb Weight	1832 kg
Weight to Power	12.0 kg/kW

CAPACITIES AND EQUIPMENT

Fuel Tank	90.9 litres
Cooling System	18.0 litres
Engine Sump	6.8 litres
Battery	12V 66AH
Alternator	75 Amps

CHECKLIST

Alloy Wheels	Yes
Adjustable Steering	Yes
Air-conditioning	Yes
Carpets	Yes
Central door locking	Yes
Clock	(digital LED) Yes
Intermittent Wipers	Yes
Laminated Screen	Yes
Petrol-filler lock	Yes
Power Steering	Yes
Power Windows	Yes
Radio	Yes
Tape Player	Yes
Rear-window Wiper	No
Remote outside mirror adjustment	(two, electric) Yes
Sun Roof	Yes
Tachometer	Yes
Cruise Control	Yes
Trip Computer	Yes

FUEL CONSUMPTION

Average for test	18.1 litres/100 km
AS2077 City/Highway	15.0/10.5 litres/100 km

ACCELERATION

0-60 km/h	5.21 seconds
0-80 km/h	7.88 seconds
0-100 km/h	11.94 seconds
0-110 km/h	14.10 seconds
0-120 km/h	16.54 seconds
Standing 400 Metres	(125.3 km/h) 18.06 seconds
LIST PRICE	$62,600
PRICE AS TESTED	$62,600

Includes options: none

Improving a Jaguar is a bit like rewriting Shakespeare. It's difficult to argue with the soundness of the basic article, but while some feel it could stand an update, others might regard a slightly dated air as basic charm. It must, then, come down to a simple matter of taste.

Converting such a product, however, rather suggests that there was room for improvement. Making an individual expression of an owner's taste is something quite different. The very fact that there are so few conversions for the Jaguar marque must tell us something, especially given the Coventry manufacturer's extensive racing pedigree dating as far back as the '50s and '60s – and, more recently, with Walkinshaw and the XJS in Europe, and with the World Sportscar Championship XJR6s which are competing this year. Jaguar's sales pitch has been one of luxury-with-performance, but it's only when you drive a 7-Series BMW or a big Mercedes that you realise that the British chassis lacks the outright agility of the German cars, and the very cosseting that the driver feels in the Jaguar is actually distancing him or her from the activity of motoring quickly. Then you also realise that The Cat's once clear-cut superiority in areas of wind noise and ride quality no longer exists either. It's now a pretty close thing. The Germans are not far behind.

All this does not matter to the Americans. They have billiard-table boulevards and a 55 mph speed limit, and they buy more of our Jaguars than we or anyone else. Economics probably dictate

BELLING

It may seem unnecessary to change Jaguar's XJ saloon can adapt Coventry's classics to suit British

Photography b

THE CAT...

...it, as Mark Hales explains, either TWR or Chasseur ...ther than American, tastes – at a price

...er Burn

that we must once more make do with what Uncle Sam has ordered.

The two variations on the feline theme reviewed here, by Tom Walkinshaw Racing Jaguar Sport, and by Chasseur Developments, are very different in degree, while broadly similar in concept. Walkinshaw has used his considerable racing success and the experience gained from it, to sharpen up the XJ saloon, and make it more of a sporting car without sacrificing any refinement. The optional aerodynamic adornments are thoroughly wind tunnel tested. Style was an important consideration, but second to efficiency. Chasseur, on the other hand, have made a statement, altering the lines and appeal of the Jaguar saloon, while acknowledging the essential parentage of the marque. Suspension and aerodynamics have received attention, but in a less subtle fashion than on the TWR machine. So TWR and Chasseur are probably only Anglicising the Jaguar rather than improving it, although it's difficult to argue that both cars are not considerably more agile in modified form.

Converted cars can be something of a mixed blessing. Often they leave us with the firm impression that the factory has spent its development budget to good effect, and that messing about with the results of that research has only upset a careful balance. This is certainly not the case with the TWR modifications. The XJ6 suspension makes extensive use of soft, voided rubber mountings in the quest for greater refinement, and the

TWR engineers have replaced "selected items" in this department, notably the rear radius arm bushes, with harder materials. There are also some additional steel packing plates on the rubber rack mountings to restrict sideways movement of the steering rack. A glance under the bonnet while someone wrestles with the steering wheel will show the whole assembly moving from side to side, and cutting down this movement obviously gives a crisper turn-in. TWR go a little further in the steering department, and for a surprisingly modest £195 will strip the Adwest power steering and equip it with a modified pinion and valve which together reduce the amount of power assistance. It's a partial and desirable answer to the oft-heard moan that the Jaguar's steering is too light, but it is still not the complete solution. Although weightier, the steering still lacks that supreme informative quality that comes from the geometry of the suspension allied to a sophisticated assistance of the rack.

Much of the problem here would seem to stem from the fact that Jaguar always have, and still do, consider the steering to be perfectly acceptable. Only recently have they grudgingly added weight to the steering of the XJ series, but only in response to continued complaints from owners.

The rear suspension also receives a touch of additional negative camber. The traditional Jaguar driveshaft-cum-top wishbone, with equal-length bottom track control arm setup, gives positive camber equal to the degree of roll which effectively lifts part of the tyre from the road surface. The problem is worse when the tyre is wide, so the best defence is to start with the top of the wheel leant inwards – negative camber. During cornering, when maximum tyre contact is needed, the wheel should be somewhere near upright. Spring rates, though, remain standard, and TWR have concentrated on the development of specially rated gas-filled Bilstein shock absorbers with a much stronger rebound control, which tie down the chassis but retain some of the ride quality.

Chasseur's approach, ironically, is more the racer's answer. Shorter stiffer springs are fitted, with gas-filled dampers – Spax, this time – and the rack receives the obvious restricting treatment, only with completely new mounts made from Teflon. The Spax dampers have adjustable spring platforms and the ride height of the car can be adjusted to suit taste or terrain, but Chasseur's suspension modifications stop at that. Attention then turns to the bodywork.

If you had gone to Kidlington for TWR's modifications then, so far, you would have spent £760 on the dampers and suspension kit, £80 for a beefier rim on the TWR steering wheel, and £159 for the power steering alterations. A total of £999, plus VAT of course, and a further £189 for fitting at the works. Bank manager still willing, you can then have a front apron, boot spoiler, and sill extensions, for a further £995 – or £1380 if you want them fitted, and painted two-tone. Special 8 × 16 inch TWR aluminium alloy wheels shod with 225/VR Goodyear tyres will fit under the standard arches and relieve you of a further £1664. To save you resorting to your calculator, the total – including the Government's share, some locking wheelnuts and a badge or two – is £4896.64 fitted.

What then of the Chasseur? This is a different approach altogether. The proprietors of Chasseur – Chas and Phil Whitaker – make no attempt to itemise their conversion because they do not wish to offer anything other than a complete car. The alterations centre more on the bodywork, and the brothers claim that maintenance of their strict standards of quality control demand that they should oversee the conversion from start to finish. It is indeed a mark of respect for the quality of their work that you need to look quite hard to see where some of the modifications have been made, even though the overall effect creates a very different animal.

Close inspection indeed reveals that this is certainly no add-on kit. The work takes eight weeks and involves a fair degree of surgery. The whole car is stripped of glass, removable panels and trim. The wheel arches and sill ends are then cut away to clear the Revolution 8 × 15 inch aluminium alloy wheels and, by current standards, the high-profile BF Goodrich 235/60 VR tyres.

The metal is then rewelded to make a watertight seam and the flares are carefully laminated on to bared metal and further sealed underneath, with a smooth skin which adds strength and prevents the ingress of water in the years to follow. While they are about it, all spot weld evidence is expunged from under the bonnet, boot and sill tops and everything, everywhere, is smoothed and cleaned up. Front and rear aprons are attached, together with the sill extensions, and the whole car is resprayed and lacquered – with similar attention to detail.

If it sounds obsessive, then it is: but there is no doubting the quality of the work. Final touches

feature they tried to avoid, provided a hard-to-ignore uplift reduction of 36 per cent, and was retained. Add to this a reduction in Cd from the standard car's 0.46 to 0.375 (TWR's figures) and it seems money well spent.

Money could well be the deciding factor overall. For the sake of argument, we'll start with a new XJ6 at £16,995. Even in 1986, it is just possible to forgive the three-speed GM transmission, controls which are indiosyncratic rather than practical, and instruments that are traditional rather than lucid, because above all it *is* still a Jaguar. The name spells something uniquely British, a detail fast becoming extinct in a hard commercial world. It can also begin to sound like an excuse when the price of the bottom of the range XJ6 is hiked to £21,892 (TWR) or £27,600 (Chasseur). Then the head-scratching must start and, as always, you have to look at what else your company's tax avoidance will buy. A BMW 735i costs £20,960, the SE version with all the extras, £26,450. Both feature niceties like a four-speed automatic box of remarkable sensitivity, superb power steering and, of course, anti-lock brakes. A Mercedes 300 starts at around £17,800, which leaves room for all the Merc add-on-essential catches and – once again – you get the four-speed box, ABS, and so on.

Between the two Jaguars, it soon becomes apparent that they don't really compare. Both turn heads (the Chasseur in no uncertain fashion) and as professionally-modified Jaguars both are unusual and fairly exclusive. The emotive argument, for me, is an easy one. I would probably have a six-cylinder Jaguar, and perhaps persuade Mr Walkinshaw to fit a manual gearbox as well as his suspension kit. If I were an accountant, the argument would be more difficult. In engineering "value for money" terms, the Teutonic alternatives look very persuasive but there must also be a band of faithful customers for whom the Jaguar is an automatic choice, irrespective of dynamic shortcomings. For these, the TWR conversion must have come as a breath of engineering fresh air.

Simply by providing some British alternatives to the Germans, TWR and Chasseur will have assured themselves some sales. Of the two, you pays your money. . . .

include much use of satin-black finish in place of chrome round the front of the car and on the door handles (£1095 will turn all chrome to black at TWR) and there are even specially-made colour-keyed door mirrors and a straight-through exhaust system which exits through the rear valance.

The bad news is that the conversion will set you back a whopping £10,600 including VAT. That's a figure which would leave you with a little change after you had driven off in a new Capri injection, but the quality of the work is outstanding, even if the aesthetics are purely a matter of taste.

On the road, the two cars are surprisingly different. Leaving aside the engines (V12/Chasseur; straight six/TWR) both of which are unmodified as tested, the cars feel surprisingly different. The Chasseur has become very much the racer of the two. The suspension is noticeably stiffer and body roll is much reduced, with grip from the fat Goodrich tyres absolutely amazing on such a heavy car. Pressing on through roundabouts will have you sliding off the gentleman's club leather seat before mounting degrees of understeer become unmanageable, and it takes a very determined effort to provoke anything like oversteer, at least on the public highways. The trade-off is a substantial increase in bump thump and general disturbance to the cabin from undulations, and a heightening of rattles and clatters though the new rack bushes. Turn-in is much improved and the car has an agility which belies its size, but although the extra rubber on the road has increased the steering weight there still isn't really enough information from the rim to know *exactly* what the front wheels are doing. The handling of the car has been considerably sharpened and the grip improved considerably, but at the expense of some refinement.

The Walkinshaw car is a good deal more subtle, both visually and dynamically. It has nothing like the grip of Chasseur, but the more extensive suspension treatment shows in the ride quality and steering feel. TWR's engineers reckon that attention to the rear suspension is the tweak that gives the big car its amazing turn-in, truly astonishing for a big saloon, and perhaps almost a little too nervous for the non-enthusiast driver. The body has lost the standard car's at times rather floaty sensation as well, and is tauter in feel all round – proof of the extensive damper development between TWR and Bilstein.

Both cars are equally stable at high speeds, despite the absence of any boot adornment on the Chasseur, but TWR substantiate their additions with some MIRA wind tunnel figures. The front apron cuts lift by a claimed 58 per cent and the boot spoiler, a

The Ford Cargo truck was hogging some of our road as it rounded the right-hand hairpin ahead. A simple flick of lock reset the XJ6's course a couple of feet closer to the verge and another turned it smartly into the bend. The delivery truck careered past, smothering the Jaguar with a sooty wake. Beyond it was a greater test of the Jaguar's new-found agility. The Cortina's rusty nose bobbed sharply to a halt across the centre line, its driver realising he was not going to make his right turn without collecting a couple of tons of luxury saloon. A deft left-right saw the Jaguar's flanks round its jutting nose with welcome composure.

The standard XJ6 is not noted for its alacrity, preferring to absorb directness for the benefit of ride refinement. The TWR chassis modifications do not exclude the car's praiseworthy compromises, they merely take up some of the slack in its responses. The impromptu display impressed both myself and the TWR man in the passenger seat who was calculating, momentarily, whether the Kidlington workshop could beat out and respray the car before its 6 pm delivery to a TWR dealer in Leamington Spa.

The Cranberry Red 4.2 Series III was to Code 2 specification – the lower of two build levels – and thus fitted with a cocktail of features from the TWR options list. Half the £6000 spent was employed announcing the Jaguar's ability, the other half in achieving it. The extensive body kit, beautifully fitted and finished, costs almost £2000. Though the flowing Series III lines do not lend themselves to adornment, the kit is conservatively proportioned, blends in well, and positively influences aerodynamics. For a little extra it can be painted in a contrasting colour, but I would have restricted exterior modification to the conversion of all-chrome to matt black which is expensive, at £1200, but enhances the XJ6's handsome profile.

The "Coke bottle cap" alloys filling the wheel arches are another visual plus-point and their large dimensions serve the car's practical development. Fat 225/55 Goodyear NCT Eagles mount snugly on to the 8 × 16 inch rims (£1700/set of four) and allow TWR's larger brake disc kit to be fitted. But what transforms the XJ6 from chauffeur-driven to owner-driven is the £1000 suspension kit comprising stiffer bushes and a set of replacement dampers to gently but firmly control the suspension's movements. There is another element, too – probably the most cost effective device you can fit to a Jaguar: a revised valve for the steering rack's power assistance. At around £200 it transforms what was light and lifeless into meaty and communicative, and the welcome dose of feedback provides ample encouragement to explore the chassis tuning.

Installed in the driving seat, only the thick-rimmed steering wheel indicates that the car is not standard, but within 100 yards it is plain that it will respond with greater enthusiasm to a sporting driving style. The first set of fast kinks delivers the steering's promise; the weighty Jaguar seems to have shed a couple of hundred pounds and changes tack with inspiring eagerness while retaining the security which is inherent with high inertia. The hard-driven Escort Turbo in front is getting in the way, but there is no chance to overtake. The 32,000-mile 4.2 under the bonnet has been doing its best to keep up through the bends: it labours under the car's weight on the short Cotswold straights.

The Escort driver scuttles off ahead, no doubt amused that such a purposeful-looking Jaguar has not got the wallop to come steaming past. Still, there is plenty of time to appreciate suspension control over the peaks and hollows of the tarmac rushing beneath the wheels. The TWR XJ6 feels little different from the standard item; crests see the car making much of its suspension travel, but when it settles back there is less squash. Dips, too, show the chassis' bias to less compliance. The tighter control is

The wheel is a clue to the transformation effected by the handling kit. Forget the German marques, a TWR is what the smart set should be driving and discerning Beasties should be wearing. £10 a throw

felt more by the driver than the passengers, who will probably not notice that the car can tackle a given road some 10 to 15 mph faster yet with the same, absorbing ride.

Road noise is slightly greater because of the wider tyres, but there is no great penalty for the TWR car's more spirited handling. Turn-in is accurate and faithful, the Jaguar never seeming short of grip; it is soon obvious that it is the engine and automatic gearbox that are holding the car back. Given that most of the XJ6's qualities are left intact, the TWR handling kit is worthy of serious consideration by those who enjoy using their cars on country roads as well as up the motorway, and hanker after crisper response. More power? No problem. That is where Code 1 specification comes in, though you do have to start with the 5.3 engine.

Most of TWR's Series III conversions are on V12 cars, and such owners have an even greater range of modifications open to them. They match the car's performance to the high level of handling achieved by the Code 2 kit and thus endow them with the performance suggested by the body styling. For starters there is a Quick Shift automatic gearbox kit which stirs the normally relaxed GM300 'box into full throttle up-changes in all gears and makes it more responsive to kick-down requests.

For even greater control, TWR offer a five-speed ZF manual 'box – and for those not happy with under 300 bhp there is a 380 bhp 6.0-litre V12 engine with more than 400 lb ft of torque. The increase in volume is achieved by a specially-made, longer-throw crankshaft and replacement forged alloy pistons and it will set you back £7500. For a more modest power increase you can have a high efficiency exhaust, worth about a 10 per cent power increase. These mods are available individually, but TWR recommend that they should be accompanied by their uprated brakes.

The 16 TWR Jaguar dealerships around the country can carry out most of the modifications offered, and cars go back to the Oxford base for major work such as engines and gearboxes. Their most popular sellers are wheels, body kits and the revised power-assisted steering valves.

Though the prices appear high, all components are warranted and careful selection from the options list can build a bargain performance car. Take a year-old V12, add the suspension kit, p.a.s. valve and wheels and tyres. Slot in the 6.0-litre engine and exhaust, quick-shift auto-box and, of course, the uprated brake kit and you have a highly desirable and effectively brand-new driving machine for under £35,000. For comparisons, just transpose the first two figures for the cost of BMW's 750i!

The body kit is a matter of taste, in the same mould as Alpina or AMG styling kits, but the simple de-chrome is remarkably effective, taking years off the Series III. TWR can claim to be the alternative to the German styling houses. They can be contacted at TWR Jaguar Sport, 15B Station Field Industrial Estate, Kidlington, Oxford OX5 1JD. (Tel: (08675) 71555.) M

JAGUAR XJ6 & VANDEN PLAS SERIES III

IT'S BEEN ALMOST 20 years since the Series I XJ6 was introduced, and, while the Series II and III cars were improved and subtly restyled in 1973 and 1979, respectively, the basic ingredients of the car have remained the same. It has always been a svelte, sensuous 4-door sedan with lots of burled walnut trim, leather upholstery and wonderful road manners, powered by the venerable but smooth XK-series dohc inline-6 engine. It has also been a car whose classic lines whispered quietly of Old Money and of a willingness to put up with a little (sometimes a lot of) inconvenience in the name of refined taste; the luxury car bought by people for whom a Mercedes-Benz is too obvious a choice.

Now, after all these years, the XJ6 is going away, to be replaced by an all-new model of the same name but virtually no common parts. The old XJ6 will go on being produced in V-12 form for the UK, German and Canadian markets for a few more years, but the last federalized 4.2-liter version for U.S. consumption will roll off the lines at the end of 1986.

The decision to replace this car was not taken lightly at Jaguar. The XJ6 has been by far the company's most popular model and, thanks to the push for better quality initiated by Jaguar chief John Egan in 1980, sales of the model are better than ever. The replacement XJ6 will probably be a better car in nearly every way, but some loyal marque enthusiasts will probably conclude that an element of old-fashioned elegance has been lost in the new car. For those people, this is the last chance to buy a Series III XJ6. The new car arrives this spring.

What do you get for your $36,300 to $40,100 in this last of the line? At the lower price you get the standard XJ6, though that standard is a little higher than in lesser cars. The XJ6 Series III comes with the usual silky-smooth 176-bhp fuel injected 4.2-liter dohc inline-6 mated to a 3-speed automatic transmission, fully independent suspension that, as only Jaguar can do it, sets the ride standard for the rest of the world's luxury cars, hand-sewn seats upholstered in soft leather, dash and door panels—and now a center console—of highly polished burled walnut, wool carpets and an onboard trip computer.

Both models also get Jaguar's new Clear Over Base paint process, in which a final clear protective coat of paint is sprayed over the base color for improved durability of shine.

In addition to all the standard XJ6 features, the Vanden Plas has its own "armchair" style seats, special door panels with inlaid burl walnut trim, individual reading lamps and magazine pockets for rear-seat passengers, custom throw rugs over the fitted wool carpeting, rear-seat headrests and upholstery in Magnolia or Doeskin leather. Both the XJ6 and the Vanden Plas are covered by Jaguar's 36-month, 36,000-mile warranty.

SPECIFICATIONS

Base price, base model	$36,300	Fuel capacity, U.S. gal.	23.8	Transmission	3A
Country of origin	England	Fuel economy (EPA), mpg:		Final-drive ratio	2.88:1
Body/seats	4D/5	Federal	15	Suspension, f/r	ind/ind
Layout	F/R	California	15	Brakes, f/r	disc/disc
Wheelbase, in.	113.0	Engine	dohc inline-6	Tires	215/70VR-15
Track, f/r	58.3/58.9	Bore x stroke, mm	92.1 x 106.0	Steering type	rack & pinion (p)
Length	199.6	Displacement, cc	4235	Turning circle, ft	38.0
Width	69.6	Compression ratio	8.1:1	Turns, lock-to-lock	2.7
Height	52.8	Bhp @ rpm, net	176 @ 4750		
Curb weight, lb	4055	Torque @ rpm, lb-ft	219 @ 2500		